DIY Instruments for Amateur Space

Sandy Antunes

Beijing • Cambridge • Farnham • Köln • Sebastopol • Tokyo

DIY Instruments for Amateur Space

by Sandy Antunes

Printed in the United States of America.

Published by O'Reilly Media, Inc., 1005 Gravenstein Highway North, Sebastopol, CA 95472.

O'Reilly books may be purchased for educational, business, or sales promotional use. Online editions are also available for most titles (*http://my.safaribooksonline.com*). For more information, contact our corporate/institutional sales department: 800-998-9938 or *corporate@oreilly.com*.

Editor: Brian Jepson
Production Editor: Rachel Steely
Proofreader: Rachel Steely
Cover Designer: Karen Montgomery
Interior Designer: David Futato
Illustrator: Rebecca Demarest

March 2013: First Edition

Revision History for the First Edition:

2012-03-26: First release

See *http://oreilly.com/catalog/errata.csp?isbn=9781449310646* for release details.

ISBN: 978-1-449-31064-6

[LSI]

Contents

Preface

What can you measure and what are your limits when orbiting in space? Learn about what physical quantities you can measure and how to design and parameterize your sensor loadout. Learn to go beyond just flying a camera and optimize your mission goals. Explore what you can play with using your own personal satellite.

Build a sensor, and you'll own a slice of the universe. Were we to describe or define a state-of-the-art sensor for a picosatellite, you could probably find a weblink to a team already launching something way cooler. Instead of listing specific gear, then, we'll work out how you can design your own sensor loadout. Knowing how to do this will future-proof your work and let you keep advancing the picosatellite field.

That doesn't mean we won't take a peek at some of the way cool sensors available currently. We'll also look at the protocols for wiring up sensors, covering I2C, TTL, Serial, and SPI; analog and digital; and a host of other concerns.

We'll examine the types of remote sensing possible and the bandwidth required. In this book, we'll also look at mechanical structures and technology efforts that CubeSats can explore.

Why Sensors on Satellites Make Sense

Ever wonder why astronomers loft telescopes higher and higher, to mountains and via balloons and satellites? We put telescopes up on mountains, launch balloons, send rockets into space and point them outward. The Earth observing (EO) folks take similar gear, fly it up high, then point it back to Earth.

Astronomers try to get our telescopes higher. Now, the main goal of a telescope is to use mirrors or lenses to essentially make a bigger eye. It's not that telescopes magnify; it's that they gather more light. In short, they make faint stuff visible. Radio telescopes extend our eye because they look at wavelengths that our eyes can't pick up at all.

A small mirror has less light gathering than a bigger one, but up high, it's less messed up by our atmosphere. From the ground, we're looking through a mile high column of air: it's like looking through a swimming pool, and the shifting air ruins our "seeing."

"Seeing" is actually a technical term. We astronomers have a knack for naming things. For instance, the big array of radio telescopes we made is called the VLA, and that stands for Very Large Array. The big explosion that started the universe we all know as the Big Bang. We tend to be very straightforward and you can see why I often say that anyone can be an astronomer if he wants.

We put stuff up to get past the wet blanket we call our atmosphere. A smaller scope higher up will outperform a larger ground scope. Telescopes are on mountains to get above the atmosphere, and also to get above the weather, because our atmosphere has a lot of weather. We also want to isolate them from city light. All the lighting of modern civilization creates what we astronomers call light pollution.

Light pollution is that orange glow that you can see above Washington, DC every night instead of stars. It makes for a really pretty sunset but it makes for really lousy observing. If you put the scope higher up, you escape that.

For Earth observing, we don't want to escape the Earth's atmosphere so much as a) observe it and b) get a bigger view. If you need to see out of a forest, you'd climb a tree. To see the Earth, to see the top of the atmosphere, or to peer at different layers, you need to do flyovers. The best way to consistently fly over the Earth and survey it is via satellites. As a bonus, you can choose specific instruments with specific wavelength regions so you can examine specific layers of the atmosphere or look for specific phenomena on the Earth's surface.

High altitude balloons are a way to go higher than a mountain, and you can get up cheaply and quickly. Some students in Spain actually lifted an ordinary digital camera 20 miles up and they did on a $100 budget. NASA's upcoming SuperPressure Balloon can lift 1,000 pounds for 100 days. It's the same concept: get things higher.

We have better visibility, less weather, longer nights, light you can't see from the ground, and a great field of view. Satellites (see Figure P-1) provide the best of these—but at a high cost. They get us higher and for a longer time, but they are a bit more expensive. And they do let us see X-rays and ultraviolet light, and give us vantage points from somewhere other than the Earth.

Given all this, we are still limited in what we can do. The "ideal detector" cannot exist, and you'll always have to make tradeoffs between capability and bandwidth, between performance and cost, and between capturing a wide view and seeing the little details. We're here to help you design your instrument mission.

Figure P-1. *GOES image of Hurricane Katrina, image courtesy of NASA/NOAA*

Conventions Used in This Book

The following typographical conventions are used in this book:

Italic
: Indicates new terms, URLs, email addresses, filenames, and file extensions.

`Constant width`
: Used for program listings, as well as within paragraphs to refer to program elements such as variable or function names, databases, data types, environment variables, statements, and keywords.

`Constant width bold`
: Shows commands or other text that should be typed literally by the user.

`Constant width italic`
: Shows text that should be replaced with user-supplied values or by values determined by context.

This icon signifies a tip, suggestion, or general note.

This icon indicates a warning or caution.

Using Code Examples

This book is here to help you get your job done. In general, if this book includes code examples, you may use the code in this book in your programs and documentation. You do not need to contact us for permission unless you're reproducing a significant portion of the code. For example, writing a program that uses several chunks of code from this book does not require permission. Selling or distributing a CD-ROM of examples from MAKE's books does require permission. Answering a question by citing this book and quoting example code does not require permission. Incorporating a significant amount of example code from this book into your product's documentation does require permission.

We appreciate, but do not require, attribution. An attribution usually includes the title, author, publisher, and ISBN. For example: "*DIY Instruments for Amateur Space* by Sandy Antunes (MAKE). Copyright 2013 Sandy Antunes, 978-1-4493-1064-6."

If you feel your use of code examples falls outside fair use or the permission given above, feel free to contact us at *permissions@oreilly.com*.

Safari® Books Online

Safari Books Online (*www.safaribooksonline.com*) is an on-demand digital library that delivers expert content in both book and video form from the world's leading authors in technology and business.

Technology professionals, software developers, web designers, and business and creative professionals use Safari Books Online as their primary resource for research, problem solving, learning, and certification training.

Safari Books Online offers a range of product mixes and pricing programs for organizations, government agencies, and individuals. Subscribers have access to thousands of books, training videos, and prepublication manuscripts in one fully searchable database from publishers like Maker Media, O'Reilly Media, Prentice Hall Professional, Addison-Wesley Professional, Microsoft Press, Sams, Que, Peachpit Press, Focal Press, Cisco Press, John Wiley & Sons, Syngress, Morgan Kaufmann, IBM Redbooks, Packt, Adobe Press, FT Press, Apress, Manning, New Riders, McGraw-Hill, Jones & Bartlett, Course Technology, and dozens more. For more information about Safari Books Online, please visit us online.

How to Contact Us

Please address comments and questions concerning this book to the publisher:

Maker Media, Inc.
1005 Gravenstein Highway North
Sebastopol, CA 95472
800-998-9938 (in the United States or Canada)
707-829-0515 (international or local)
707-829-0104 (fax)

We have a web page for this book, where we list errata, examples, and any additional information. You can access this page at:

We have a web page for this book, where we list errata, examples, and any additional information. You can access this page at *http://oreil.ly/DIY_am_space*.

To comment or ask technical questions about this book, send email to *bookquestions@oreilly.com*.

For more information about our publications, events, DIY kits, and products, see our website at *http://www.makermedia.com*.

Find us on Facebook: *http://facebook.com/makemagazine*

Follow us on Twitter: *http://twitter.com/make*

Watch us on YouTube: *http://www.youtube.com/makemagazine*

Add us on Google+: *http://plus.google.com/+MAKE*

Acknowledgments

The author would like to thank the patience of his family in surviving yet another book. We also thank those students in the Capitol College instrumentation and remote sensing courses for their valuable feedback in the presentation of much of the material in this book. Finally, a thanks to *Science20.com* for being the first to believe in the Project Calliope mission.

1/Understanding Measurement

Anything can be measured. A sensor is a device for measuring quantity. *Sensor* and *detector* are fairly synonymous. An instrument is one or more sensors. A telescope is one or more instruments coaligned, often with a focusing device.

Do you know what a creepmeter measures? Measurement is the heart of science. What distinguishes science from opinion or philosophy is measurables. The root of science is facts that are determined by actual observation, compared, then extended into predictions.

Any good measurement has three parts: the number value, the units you're using, and the error. If I say I am 6 feet tall, that's a number (6) and a unit (feet), with a presumed error of *within an inch or two*. All three parts are crucial.

What can you measure on the Sun? We often think of the Sun as a big glowing yellow ball in the sky Figure 1-1, but the Sun is a complicated entity. Think about everything you might be able to measure. We'll get back to it in a moment.

A peek at the Wikipedia list of gadgets (*http://bit.ly/YCY5gL*) reveals the insane number of things you can measure and devices you can use to do it. Here's just the list from A to C:

accelerometer
: acceleration

actinometer
: heating power of sunlight

alcoholometer
: alcoholic strength of liquids

altimeter
: altitude

ammeter
: electric current

anemometer
: windspeed

Figure 1-1. *The Sun during the day*

atmometer
: rate of evaporation

audiometer
: hearing

barkometer
: tanning liquors used in tanning leather

barometer
: air pressure

bettsometer
: integrity of fabric coverings on aircraft

bevameter
: mechanical properties of soil

bolometer
electromagnetic radiation

calorimeter
heat of chemical reactions

cathetometer
vertical distances

ceilometer
height of a cloud base

chronometer
time

clap-o-meter
volume of applause

colorimeter
color

creepmeter
slow surface displacement of an active geologic fault in the Earth

Here are just some things can you measure about the Sun: its size (a.k.a. volume), distance from Earth), mass, and temperature. Density and surface area are worth noting here, too. Also, it's rotating, so we can measure that. Then there's light output, perhaps broken out by spectrum (amount of each *color*) all the way from radio through visible light and up to X-rays and gamma rays (Figure 1-2 shows it in ultraviolet). Particle emission, too. Temperatures, obviously. And chemical composition—what elements exist, including its metallicity—which you can consider "everything other than Hydrogen, H, and Helium, He").

The Sun has strong magnetic activity, so you'll want to measure the magnetic field and the electric field. You'll want to measure the sound waves that go through it, perhaps add some helioseismology. Opacity (how transparent parts of it are)? Rate of fusion (conversion of H to He + heat): measure that. Particle emissions, from high energy stuff to neutrinos, are measurable too. You can derive values such as its gravity (from mass) and its age (from a variety of things). The Sun is moving through the galaxy, so you can measure that.

The Sun is joyfully complex. Given the Sun has layers, we'll want to measure all these properties at different layers. The rotation is at different speeds at different distances from the equator, and in fact, most of the things you measure change with time as well as location. Add in transient activity—loops, flares, CMEs, and other brief yet potent events.

Picosatellites are an ideal platform for testing new instruments and opening up new measurement horizons.

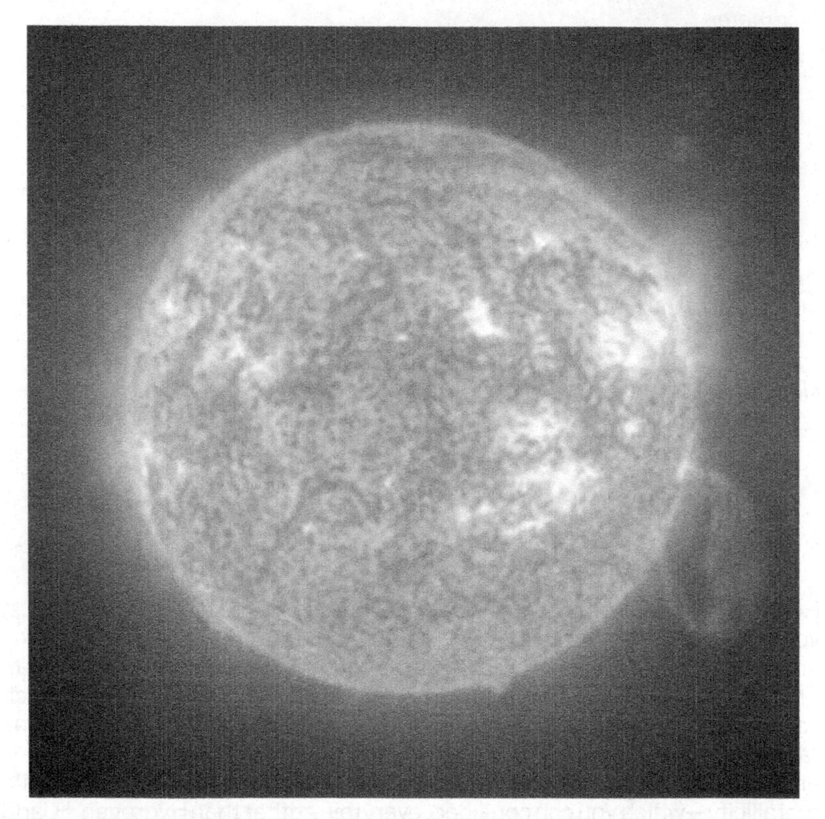

Figure 1-2. *The Sun in ultraviolet, image courtesy: NASA STEREO*

Brainstorming

As in the first book in this series, *DIY Satellite Platforms* (O'Reilly/MAKE) and satellite design in general, you will be constrained in your budgets for size, weight, mass, and money. Within those limits, you need to design a sensor loadout.

As an open-ended design qualification question, we offer these design criteria:

- Why do it?
- What data do you want?
- How will you pitch it?

The best way to understand space is to figure out how to explore it, the best way to learn engineering is to design a mission, and the best way to emphasize mathematical rigor is to apply math to solve interesting problems. The current rapid advances in picosatellite access are shifting satellite technology from the old model of "can it be built" to a newer model of whether you can come up with a concept that is worth flying.

At the brainstorming stage, pie-in-the-sky ideas are great. Picosatellites are born for trying new ideas. Your payload concept can be about an instrument or measurement, or a technology test or tech demo, or a science/hybrid mission such as sonifying the ionosphere or creating art in space. When coming up with what to fly, deliberately keep it very open to encourage imaginative as well as critical thinking.

Picosatellite payload brainstorming requires both intuitive and mathematical development of two key scientific and engineering concepts. Scientifically, you must define a useful set of items to measure in orbit. As an engineering challenge, you must balance trade-offs of power, mass, and size of your satellite versus the goals you wish to achieve. These loosely break into the categories of design and implementation.

The use and choice of measurement is what differentiates science from opinion. Fundamentally, a measurement consists of a number value, the units of measure, and the error range in the measurement. Designing a picosatellite instrument package engages the same path of analytical thinking, moving from base dictionary properties such as *space has no air* or *the Earth is blue* to scientific thinking on concepts such as orbital drag or atmospheric opacity.

Your scientific goals will provide the parameters that your detector engineering will achieve. One key component of scientific measurement is being able to distinguish between two similar objects or events. For example, Figure 1-3 shows the aurora on Earth, due to solar activity charging up our ionosphere. Figure 1-4 shows a similar aurora on Jupiter: same solar cause, different planet. While engineering can build a detector to measure a property such as *magnetic field*, science informs the engineering so that measurements of the two aurora allow you to compare and contrast their properties.

The core engineering concept is the use of key trades and resource budgets. A key trade is the balance of connected opposing quantities. Everyone faces this in using common consumer items or in managing their time. Reducing electronics weight raises the cost. Using more power-hungry devices decreases your battery lifetime. Spending time watching TV means you get less sleep. In engineering thinking, these trade-offs are quantifiable: you can put a number to the relationship and use your data to decide.

For picosatellites, we emphasize the cardinal spacecraft limiting budgets of: Mass, Size, and Power. Add too many instruments, and one or more budget

Figure 1-3. *Earth aurora viewed from the ISS, image courtesy NASA*

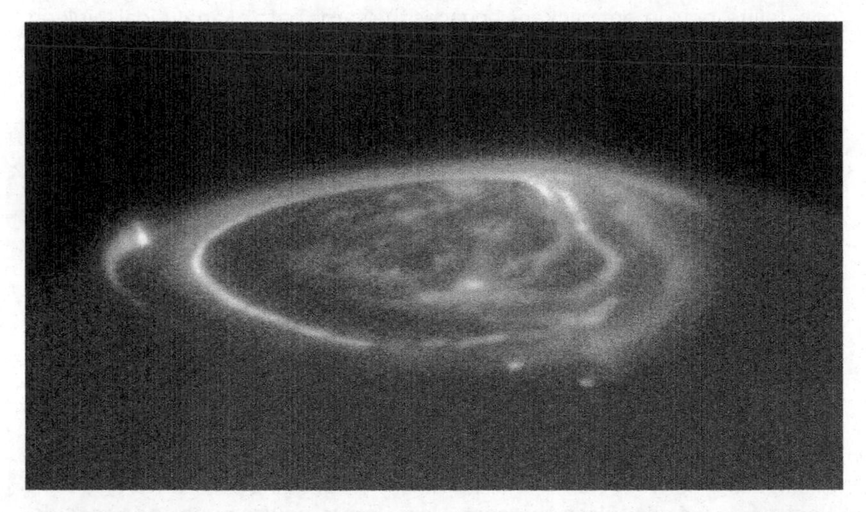

Figure 1-4. *An aurora on Jupiter, viewed by the HST, image courtesy NASA. Is this similar to Earth, or different?*

is exceeded and the mission is not viable. Fail to make good use of your budgets, and your mission may be underutilizing its capacities.

A TubeSat picosatellite lifts 200 grams of payload. That's about 7 ounces. Viewed one way, that's less than half a can of soda. But it's enough to lift an entire Nintendo DS game handheld into orbit. Two hundred grams can mean a lot of electronics.

When I committed to this project, I didn't yet have the specific electronics in mind. I've built mini guitar amps and guitar sound processors that come in well under 7 ounces. I assumed I could kit-bash stuff and create my own schematics for the final assembly. What I didn't expect was that there would be companies that already build everything I need.

You likely already know about MAKE's Arduino books, such as *Getting Started with Arduino*. Robot Electronics is a shop that also has some good tutorials. I-CubeX is a Canadian company that creates sensors for performance art, concept pieces, human-computer interaction, and just general sensor fun. Any of these has much of my payload in easy-to-assemble Lego-like fashion.

As a case study (since it was my payload), I-CubeX sensors returned MIDI. What's MIDI? MIDI is an electronic musical instrument data format. It lets you separate the instrument from the actual sound generation. The instrument—a MIDI keyboard, MIDI drum, MIDI saxophone, MIDI satellite—generates event messages that you can feed into a sound card or synthesizer. It's kind of like transmitting a player piano roll: it has no sound itself, but it lets other devices create sound.

My concept was to sonify the ionosphere, to fly a satellite with MIDI-based sensors that returned their data as instrument riffs rather than raw numbers. True, the data is still just raw numbers, but the interpretive aspect is easier for non-scientists.

As a first step in my path of instrument design, I ordered a batch of sensors just to start playing around. This is an important part of sensor design—start early, try lots of things, and don't commit to your final loadout until you've played around a bit.

For starters, I had a single magnetic sensor and a sensor-to-MIDI converter. The converter can take up to 8 channels, so if I have weight to spare, I can add more sensors. Optical, definitely. Temperature, likely. Vibration, maybe.

As a weight test, I weighed out my first batch of parts. I added in an FM transmitter (representing a to-be-designed add-on) and, of course, a box to hold them. I'm under 3 grams—well under half my weight budget.

During the brainstorm stage, feel free to fail. I tried a heliophone kit from a UK company to see about adding on-board signal processing akin to a theremin. Eventually I discarded this idea, but the toying around helped me later define how I wanted to process my signals.

My next analysis was on my power budget. The more sensors and sound processing I stick in, the more power it sucks up. We have a fixed amount of juice from the battery/solar cell setup. Limiting yourself to 3.5V and 500mA is a strong limit, but well within the capability of today's easily available robot and drone sensor parts.

The result of my brainstorming and prototyping defined the mission. My satellite would not take images, but would sample the magnetic field, the temperature, and the ambient brightness levels experienced by a satellite that travels through the ionosphere at a rate of one orbit every 90 minutes.

My satellite sensors return data formatted into MIDI. You (on Earth) take the MIDI data and you can make it sound like anything. Piano, trumpet, footsteps...just dial in a different sound and run the MIDI data and it'll play in that voice. The MIDI events give you pitch and intensity; the rest is up to the remixer.

Like techno? Assign magnetic and temperature to two harmony drones, opticals on different sample triggers, and add a drum track. Like ambient? Try magnetic on organ, temperature as a phase modulator, optical on chimes. Punk? Put everything on guitar and speed it up. Want space whales like Star Trek IV? Map everything to whale song samples. I'm doing music from space, using a 200-gram instrument in a half-kilogram case launched 312 kilometers up.

What 200 grams would you put into orbit?

2/Introduction to Instruments

There is no ultimate detector, no detector that has an infinite field of view with perfect spatial resolution taking rapid frames covering the entire spectral wavelength (plus particle events) with zero noise. We'll cover what sensors and detectors do, and how to balance the key tradeoffs required to make the best sensor for *your* task.

Let's get some definitions out of the way:

- A *sensor* is a device for measuring a quantity. *Sensor* and *detector* are fairly synonymous.
- An *instrument* is one or more sensors.
- A *telescope* is one or more instruments coaligned, often with a focusing device.

Even something as simple as "let's go out and look at the night sky" requires instrument design. If you grab binoculars, you can see bright stuff only, but it's easy to find it. In setting up a telescope, you need to choose an eyepiece to maximize the amount of light and detail seen, while also making sure the field of view is wide enough to a) let you find the object and b) see the entire object. And if you're tracking something that changes, such as a variable star, you'll have to come out many nights in a row to see any variability.

So for a simple, known problem—look up at the night sky—there are still design issues at stake. When you start with an open slate—"we can fly anything on our picosatellite"—the design problem becomes a fundamental issue that must be tackled.

Basic instrument design has to deal with detection, saturation, and variability, while also tracking the bandwidth involved.

Detection

You don't want to miss out on data because your detector just can't see that low. This is like your eyes on a very dark night—you can barely see things. This is why night vision goggles were invented for people—to raise the signal to a level your eye can get useful information out of.

Saturation

You don't want to saturate (overload) your detector by having it experience signals too large for it to measure. Think of how your eyes respond to a spotlight—poorly. Too much light means you can't see anything.

Variability

Signals can change over time, and you want to make sure your detectors capture that. In our eye example, imagine you have a flashlight you can turn on only once a minute. That gives you insufficient information to navigate a nighttime terrain at any speed. You'll also miss any passing wildlife. If you increase your light source's *blink rate* so you can see once every second, you'll get a stroboscopic effect where you can see everything clearly. If your light flickers at 24 times a second, it's essentially the same as being fully lit, and anything faster doesn't gain you much advantage. So you want to make sure what you're doing is fast enough to capture the action, but not oversampling things so that you have extra data with no real change observed. See Figure 2-1 for an illustration of timing issues.

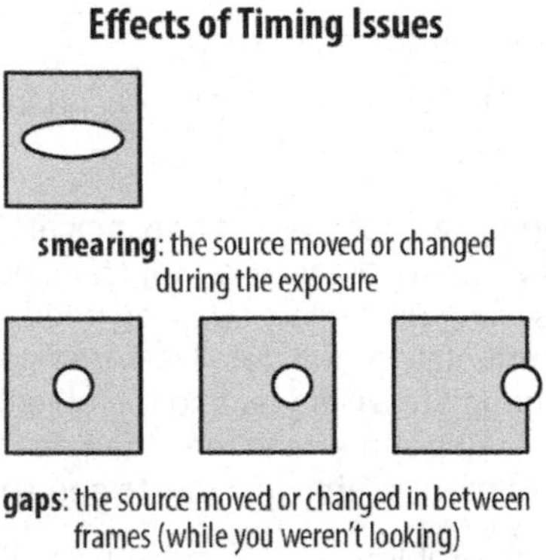

Effects of Timing Issues

smearing: the source moved or changed during the exposure

gaps: the source moved or changed in between frames (while you weren't looking)

Figure 2-1. *Why timing is a tricky subject*

Bandwidth

Finally, bandwidth is the amount of data being transmitted, and is usually a limited quantity. You can only download—and sort through—a given amount of data. For example, movies tend to run at 24 frames per second (fps). You can make a movie at 1,000 fps, but at that point you're just wasting film (or digital storage space) with overkill. On the other hand, if you are severely limited in bandwidth—say, you have to watch the movie over a slow modem connection—you have to somehow limit your data. Perhaps you take fewer frames per second, or you take smaller frames, or you compress the data and risk artifacts and jaggies, but your bandwidth needs will constrain how you capture the data.

We don't want to saturate the detectors, nor miss data, so we need to tune them to the range expected in space. We want to capture all the data and changes we can, within our bandwidth limits. And we want it to work so the data has utility. Time to go into a little sensor formalism. The five primary components we will explain are our spatial or *Imaging resolution*, our choice of color or *Spectral resolution*, our ability to discern brightnesses or *Photometric resolution*, and being able to capture events or *Timing resolution*.

Parameterizing a Mission

Not every picosatellite is a sensor mission. The quite capable list of past CubeSat missions at Wikipedia, for example, lists several technology demos (*http://bit.ly/Ys3vuK*) and operational tryouts, as well as several radio relay satellites. This book assumes you want to capture data, not just operate a satellite.

Mission Domains

One easy way to define a mission is to assess what science branch it falls into. This schema is particularly nice, as it maps to the NASA science domains and thus matches much of both the published technical literature and the grant and proposal world as well (Figure 2-2).

1. Earth-observing
2. Solar
3. Astrophysics
4. Planetary

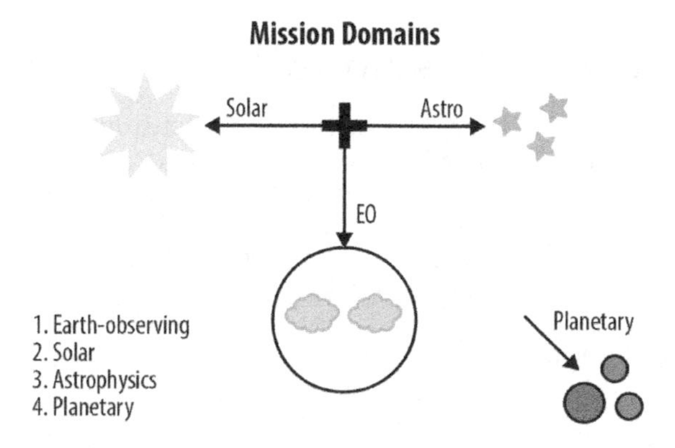

Figure 2-2. *Which domain you observe is primarily an issue of which direction you point*

Licensing

Does your satellite need a license? Space is highly regulated. Some of the regulation can be handled by your launch provider, but there are three areas you need to look into yourself. The first is communications: in the USA, either the FCC or the IARU must approve your mission's communications profile.

Second is for detectors—in the USA, NOAA (see below) approves *remote sensing*, which may or may not included non-image work as well.

If you are from the US, observing the Earth with visual (camera-like) detectors, you need to get a special waiver from the National Oceanic and Atmospheric Administration (NOAA, license info at *http://1.usa.gov/10GOLf1*). In other countries, you likely have a government agency that provides licensing. This is to prevent just anyone from being able to launch a spy satellite.

Formalities are important. NOAA states "It is unlawful for any person who is subject to the jurisdiction or control of the United States, directly or through any subsidiary or affiliate to operate a private remote sensing space system without possession of a valid license issued under the Act and the regulations."

Most picosatellites are going to have very poor spatial resolution, and therefore are not likely to be considered a threat. Therefore, getting this waiver is a question of paperwork, rather than being something to worry about during your design.

Finally, there are requirements for insurance (if your satellite damages another) and deorbiting (avoiding leaving debris) that need to be investigated. These are areas in which launch providers and existing CubeSat consortiums can assist. Alas, I am not a space lawyer, and thus will refrain from giving advice.

Look at Past Missions

Before you begin choosing your detectors, you must study what has already flown. While there is no single defined *mission specification*, I suggest the following bulleted list to help you quickly parameterize the instrument engineering and science designs of past missions. A good way to begin is by choosing one or more parameters from the first category (mission domain), one or more from the second category (type), etc.

- Mission domain
    - EO
    - Solar
    - Astro
    - Planetary

- Type
  - Imaging
  - Spectrometric
  - Photometric
  - Timing
- Pointing?
  - Pointing
  - *In situ*
  - Survey
  - Active/passive
- Wavelength regime (energy range)
  - nm
  - mm
  - m
  - eV
  - keV
- FOV
  - Extra-wide
  - Wide
  - Medium
  - Narrow
- Resolutions
  - Spatial: L/M/H
  - Spectral: L/M/H
  - Photometric: L/M/H
  - Timing: L/M/H
- Description
  - Example: A camera using a 4x4 array of CCDs

Start with the broad science goal of the mission—is it Earth-observing, astronomical (pointing outward), Orbital sensing (looking at other items in orbit), or *in situ* (measuring the environment around the satellite itself)? Did it serve a technical function such as acting as a radio relay or testing a solar sail, or was the primary goal to acquire data via sensors?

Given its mission type, what is its specific target or class of targets that it wishes to observe?

Next, look at the wavelength regime the mission is using. Do the sensors detect visible light, radio or submillimeter light, X-rays, gamma rays, particles, or materials?

Who was involved, and how big of an effort was it? This information is useful for comparing missions. A NASA Great Observatory is very different from a single-detector SMEX satellite, and the latter might be more informative when scoping out picosatellite concepts because of its reduced goal set.

What orbit was involved? Low Earth Orbit (LEO) is what we assume is typical for prospective picosatellites, but there are many orbits possible, including high Earth and geosynchronous orbits, solar orbits, even lunar and planetary orbits. Your choice is limited to what altitude your rocket can put you into, and so the orbit itself will be a clue as to the payload and sensor choices.

Given a Low Earth Orbit, you are going to loop around the Earth every 90 minutes (give or take 20 minutes, depending on the specific orbit assigned to you). Half of that might be in darkness, with the other half in sunlight. Your orbit is rigid and fixed (relatively) but the Earth is turning below you, so you will have a different part of the Earth underneath you over the course of a 24-hour cycle. If you are looking down at Earth, the region below might have different lighting each time you cover it, or (if a Sun-synchronous orbit), it might always have the same lighting. If you are pointing at an astrophysical target, the Earth may or may not block (or occult) your target for part of each orbit, and whether there is occultation depends both on the astronomical target's location and the date you are observing from orbit.

Now you can look at each detector on the missions you are analyzing. Break them into their categories (imaging/spectrometric/photometric/timing). Look at the specific hardware, be it CCD, Lidar, etc.—there are hundreds of pieces of potential detector hardware.

For each detector, get the resolution numbers. Indicate, for each detector, its range for spatial, spectral, timing and, if possible, minimum and maximum brightness.

Finally, look at the operations mode required. How many contact passes to downlink data to the ground did they have? How many will you have? Use past missions to understand how to better design your mission.

As a report, a mission analysis includes all five of these topics. We'll also expand on this format and use it later to parameterize your own mission design:

1. Summary of mission: type (EO, Astro, etc.), wavelength regime (radio, optical, X-ray, etc.), science goal (observe stars, track water, etc.), participants (NASA, JAXA, etc.), orbit (LEO, L5, etc.), and dates operational.
2. Summary of detectors involved, including the detector category (imaging, spectrometer, etc.), the specific detector type (CCD, Lidar, etc.), and how they work.
3. Detector Data resolution limits. Indicate, for each detector, its range for spatial, spectral, timing and, if possible, minimum and maximum brightness (e.g., a 128×128 imaging detector covering 1.5 degrees of wide-field sky view over a 15-250 keV wavelength range taking 16 second-long exposures, each with 4-second time resolutions, able to resolve faint objects but must avoid the Crab Nebula, Sun, and other bright objects).
4. Operational details, such as the number of contact passes and comm support required (e.g., LEO, TDRSS passes 5 times daily), the approximate daily data rate (e.g., 1TB/day), and any other interesting details (such as cryogen/consumables).
5. Why flown or summary of impact. Historically, how well did this do, how advanced was it, and what was its impact in its category (e.g., was first testbed for CCDs, first to discover black holes, predecessor to HST)?

This sort of historical analysis will let you use past missions as a template for designing your future mission. Further, this is very handy if you want to advance the state-of-the-art of picosatellites, and not just duplicate past efforts. Here is a blank table to get you started:

Mission domain	
Type	
Pointing?	
Wavelength regime (energy range)	
FOV	
Resolutions	
Description	

3/By Wavelength

The first consideration for a detector is what wavelength of light or what energy quantity you will measure. On Earth, people buying a camera already presume they are taking color pictures and do not think about this. For a space-borne instrument, however, the choice of which color—which wavelength—or which energy regime you are measuring is crucial.

Each wavelength band provides a window onto a different aspect of nature, a different set of physical processes, and a different slice of whatever your target is. Whether looking out (for astronomy, solar, and planetary sensing) or looking down (for Earth observing and surveillance work), the wavelength band you choose will strongly determine what you see.

By analogy, if you are looking for people during daylight hours, you will want to look in ordinary optical light. At night, you'll want to switch to infrared, so you can see their heat signatures even though they are *invisible* in the dark. The choice of wavelength—optical or infrared—determines which sensors you will require.

The Spectrum

We'll use the terms *wavelength* and *color* fairly interchangeably. You can also use the term *frequency* (frequency of light, not frequency of how often something occurs) and, in some cases, temperature, as a synonym for wavelength. For now, let's just use wavelength and color as synonyms. Figure 3-1 shows the electromagnetic spectrum.

Each wavelength shows light emitted by sources that have different temperatures. Cold items emit a little radio. Cool items emit infrared (IR). Room temperature and hot objects emit optical and ultraviolet (UV) light. Very hot objects or objects undergoing nuclear processes emit X-rays and gamma rays.

Just to complicate things, most items emit a *spectrum*: a range of light, rather than just a single wavelength. So a hot object emitting bright blue light will also emit some IR and UV and maybe a bit of radio. The specific light they emit also depends on what they are made of.

For example, the Sun primarily emits light in the optical range, and that is mainly in the colors of green and yellow. However, it also emits copious amounts of IR and UV, and X-rays, and lots of radio emission.

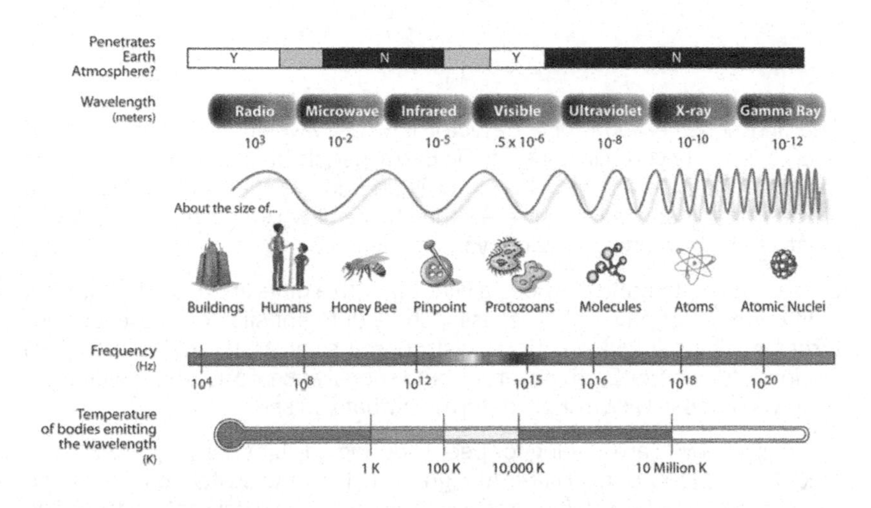

Figure 3-1. *The electromagnetic spectrum, your choice of colors. Graphic courtesy of NASA.*

Add to this non-photon emissions, and you open up a new range of energies to detect. Particle emission (charged protons and electrons, neutral neutrons) stream from the Sun or other energized sources. Electric and magnetic fields can be measured as well, adding to the picture.

For Earth Observing, each wavelength indicates a different layer or a different material. For astronomy, each color likewise indicates layers or materials, and also other factors like movement. There's a good summary of "Rainbow Astronomy" on Wikipedia (*http://bit.ly/1ONAUr6*), and we'll go through some of the use of color here.

Color is very important in Earth observing. Gas, water vapor, and dust scatter light. The dominant physics principle here is Rayleigh scattering, where blue gets scattered more than red. There is also Mie scattering, which is fairly flat and scatters everything.

As a result of this scattering, observationally, the Sun's light has less blue directly visible, since that gets preferentially scattered. And as the atmosphere gets thicker or dustier, using the list of colors in order of decreasing wavelength ROYGBIV (red, orange, yellow, green, blue, indigo, violet), eventually the green, yellow, even orange gets scattered, yielding pretty sunsets.

This means the atmosphere acts as a filter or distorter of the original Sun. Any color image taken via a picosatellite at high altitude will be bluer; we may over- or underestimate the amount of blue if we don't know where the image is taken from.

This leads to a neat benefit. If you know the light source (e.g., the Sun) and you observe the spectrum received, you can deduce the material by what was observed. Add in *nonselective scattering* by particles, which scatter only that light which has a wavelength greater than or equal to their size, and you get more color information. Nonselective scattering is why clouds are white: they scatter every visible wavelength evenly.

All real substances also scatter light. Specular reflection is like a mirror: it is the smooth and shiny bouncing of light. Diffuse reflection is how rough surfaces scatter light at a variety of angles. Substances have a mix of both types of reflection.

As a result, each substance has a *fingerprint* of how it reflects some colors, and not others, leading to a spectral reflectance curve. If you can measure the colors accurately, you can identify and differentiate one material from another. Figure 3-2 shows the characteristic color fingerprint for a leaf, overlaid against the 4 color bands with which the TRMM satellite observes it (blue, green, red, and infrared). By combining the 4 measurements, you can identify the substance.

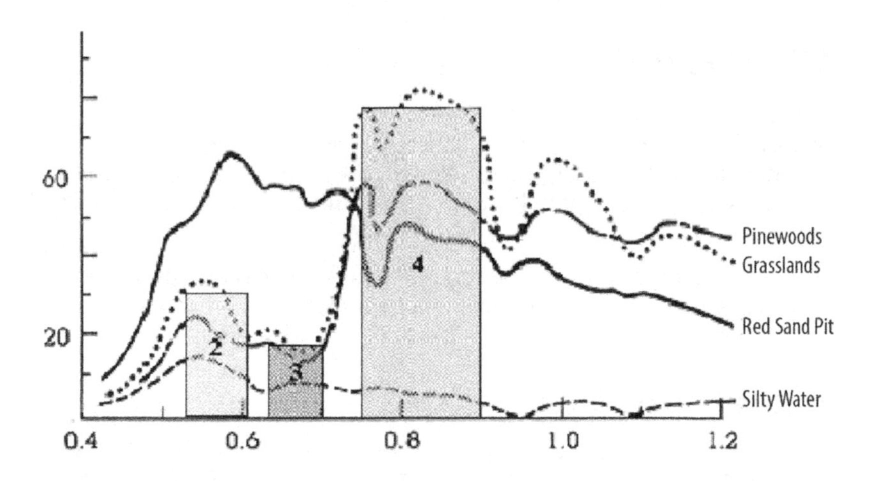

Figure 3-2. *Spectral fingerprints for several plants showing why observing in several spectral (color) bands is useful. Source: United States Department of Transportation.*

Reading Spectral Profiles

In Figure 3-2, there are spectral fingerprints for soil, vegetation, and water, showing why observing in several spectral bands (colors) is important. The green, red, and pink bars represent the 3 colors the Landsat ETM+ instrument can observe (equaling visual green, visual red, and near infrared or IR). The height of each bar indicates how efficient the detector is—as shown, it has higher sensitivity and accuracy for IR, moderate for green, and low for red. The overlaid line curves each represent the spectral profile or color mix of each material. For example, *red sand pit* is mostly greenish with a little IR while *grasslands* has some green and a large amount of IR. Viewing multiple colors lets you identify materials by their spectral *fingerprint*.

Gamma and X-rays

The high-energy wavelengths of gamma and X-rays are produced by extremely energetic events. In astrophysics, this means flares, supernovas, black holes, stars cannibalizing each other, and other very interesting activity. In Earth observing, this indicates nuclear testing and therefore is worth noting. X-rays are also emitted by radioactive materials, making their detection an excellent way to find heavier elements.

Ultraviolet

Ultraviolet (UV) light is invisible to us, primarily because our atmosphere blocks most of the Sun's UV emissions. But most astrophysical objects (the Sun included) put out a large amount of their emitted light as UV. The Hubble Space Telescope, for example, observes in both optical and UV to make its astounding images.

UV is also useful for Earth observing, as the response (absorption and scattering) of UV by the upper atmosphere to UV light can let us know some of its composition.

Violet and Blue

The color violet in images of other planets can indicate the presence of aluminum (Earth, as shown in Figure 3-3, is quite blue, as you well know). "Carbon stars" (stars that have been burning so long they have created a fair amount of carbon inside) are much fainter in violet than a typical star, so a lack of violet can indicate carbon. A glow of violet or purple can indicate scattering of light by dust.

In astronomy, blue can indicate very hot stars or stars that are metal-poor (i.e., they don't have much metal). So we get chemical information from the amount of blue we observe. Blue can also indicate a Doppler shift for an astrophysical object moving toward us. Planetary atmospheres with blue in them may have methane, since methane absorbs red and leaves the blue to shine through.

Perhaps most important for Earth observing, blue is the color of water and lets us track oceans, seas, and rivers. Lightning glows blue due to ionized atoms or excited molecules emitting blue light.

Figure 3-3. *Earth has many shades of blue. Image from space, courtesy NASA.*

Cyan and Green

Comets can have a cyan (Figure 3-4) or greenish glow in their coma and tails, indicating the presence of chemicals such as poisonous cyanogen (CN) or silicates (silicon-based compounds). Some planets (like Uranus) have methane, which leaves a blue or cyan color visible. Dust also will scatter light and can lead to a cyan tinge.

Green is also the color of vegetation and therefore crucial in Earth observing. Each type of leaf reflects a different green. The wetness or amount of water will also provide specific color *fingerprints* in the light observed. Algae blooms and coral reefs can result in green features in bodies of water, letting us track their growth. Phytoplankton blooms show green against the blue of water, leading us to track their growth and find contaminates or heat changes that led to such growth. Figures 3-5 and 3-6 show a couple examples.

Figure 3-4. *Comet image, cyan indicating comet structure. Image courtesy of NASA.*

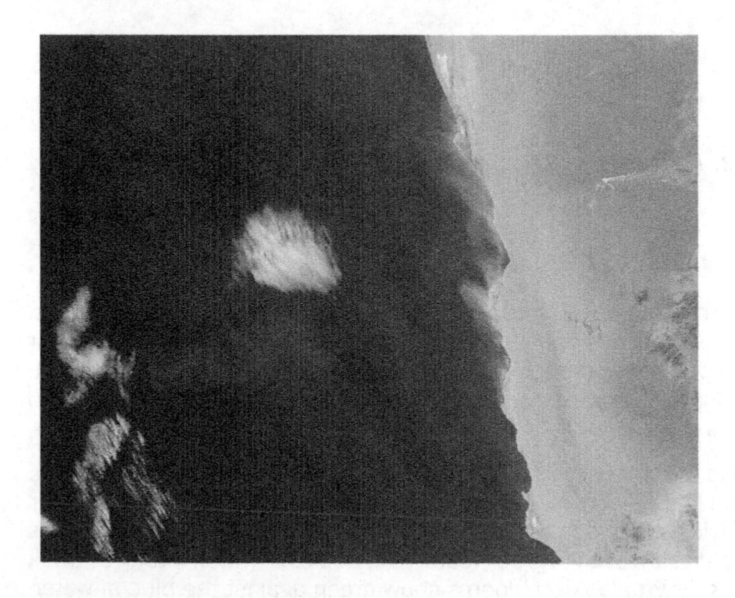

Figure 3-5. *Green indicates hydrogen sulfide emissions off the coast of Africa. Image courtesy of NASA.*

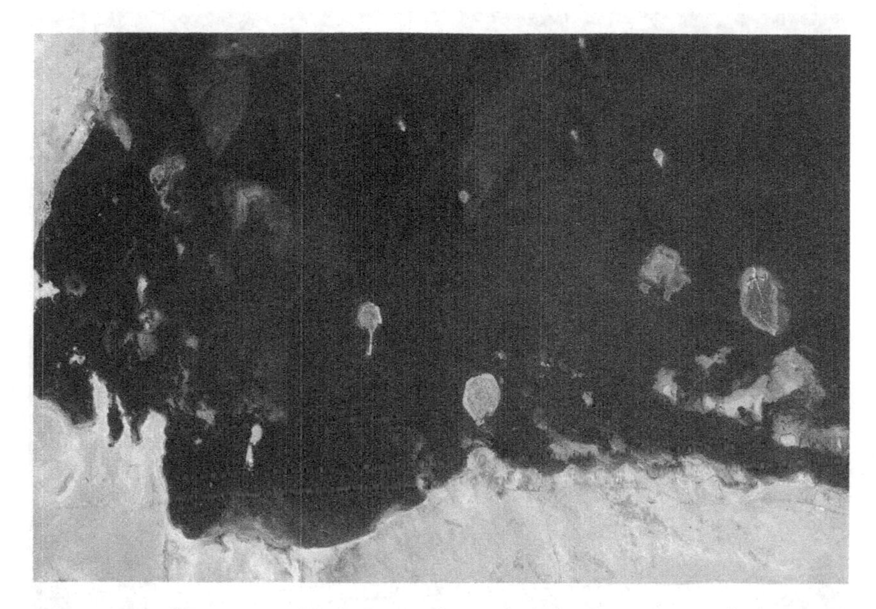

Figure 3-6. *Green can also indicate the presence of coral, as in this Persian Gulf reef. Image courtesy NASA.*

What value does such color analysis provide? An example from Wikiversity notes "The milky-green colors along Namibia's coast indicate high concentrations of sulfur and low concentrations of oxygen. Episodes like this aren't just colorful, they are actually toxic to local marine organisms. Fish die in the low-oxygen water; however, what is deadly for the fish can be good for birds that feed on their carcasses. Likewise, lobsters crawling ashore to escape the toxic seawater can make meals for locals. And some species of foraminifera —tiny, shelled marine organisms—actually thrive in the oxygen-poor sea floor sediments off the Namibian coast." A little color provides a lot of interpretation.

Yellow and Orange

We see the Sun as yellow, so certainly that is a wavelength band worth considering. Other items are also yellow: Jupiter's moon Io, for example (shown in Figure 3-7), has a surface laced with the results of volcanic activity and shows up as yellow, primarily due to sulfur. Jupiter has orange atmospheric bands, the result of either sulfur or perhaps complex hydrocarbons. Tracking these bands gives us clues to the movement of its atmosphere.

Figure 3-7. *True color image of Io. Image courtesy NASA.*

Red in astronomy can indicate a cool star or a bright object that is also moving away at high speed (as Doppler shifts change the color of fast-moving objects). Figure 3-8 shows an example of this.

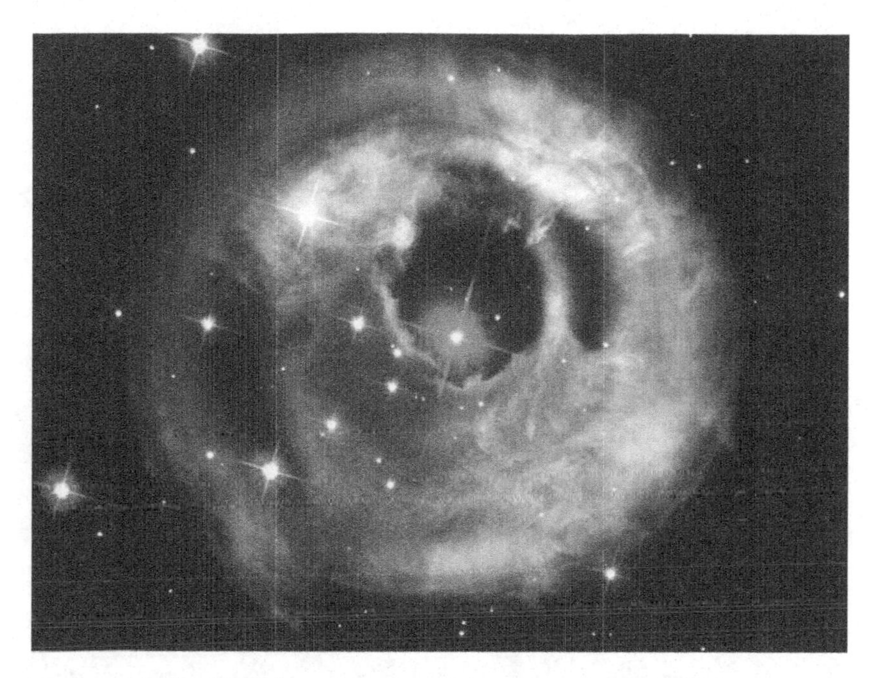

Figure 3-8. *The V838 Mon nova, in red. Image courtesy of NASA.*

Red on Earth can indicate scattering by dust—possibly kicked up by volcanoes—as well as revealing objects (like fall leaves) that may be intrinsically red.

Infrared

The easiest interpretation of infrared (IR) is that it shows us the temperature of an object (see Figure 3-9. Items that radiate heat radiate infrared photons. Whether for astronomy or Earth observing, IR lets us track temperature information, particularly for cooler objects that aren't visible in the optical spectrum. Clouds of gas or atmospheric clouds both will yield IR.

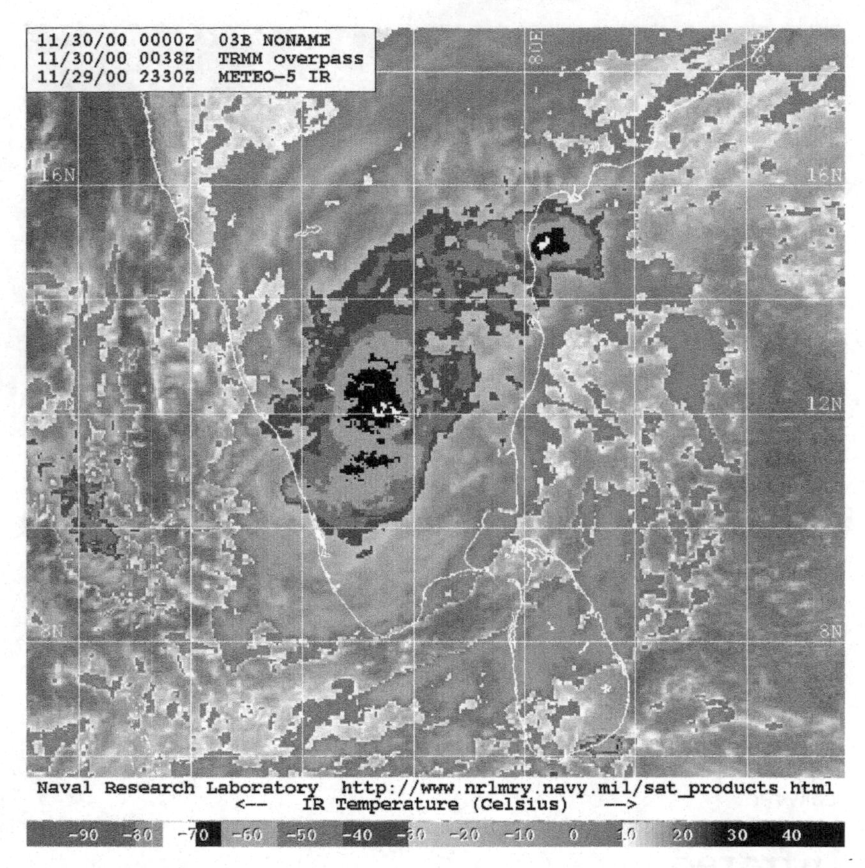

Figure 3-9. *Infrared (IR) image using the TRMM satellite, with temperature key indicating how IR light corresponds to temperature. Image courtesy of NASA TRMM mission and NRL analysis.*

Microwave, Submillimeter, and Radio

The even-longer-than-IR wavelengths move into the radio region. Radio emissions tend to come from large amounts of very cool gas, making radio a fine astrophysical tool for looking at cool hydrogen and a good Earth observing tool for examining clouds. Radio lets us complete the picture.

4/Fundamental Detector Types

As we will often note, there are four flavors or types of data:

- Images → imaging
- Brightness → photometry
- Color → spectroscopy
- Timing → variability

Each involves a separate data goal, and together all four fold into your detector profile and its bandwidth needs (Example 4-1).

Example 4-1. Examples of data types

```
Brightness: e.g. 3.846 x 1,026 W in luminosity, 4.83 absolute magnitude

Spectral: e.g. "color image plus IR/thermal"

Timing: e.g. "On during the day, off at night"
```

You can also mix these data goals. An imaging spectrometer is a camera (imaging) that gives you a separate data frame for each color. Instead of seeing one picture, you can compare, for example, the image in red against the image in blue.

A light curve is how the imaging, brightness, or color of an object changes over time. Light curves are crucial for analyzing the variability of a changing object. Light curves are often used to find asteroids or even extra-solar planets through occultation (one object moving in front of another). A brightness shift often indicates occultation.

A spectral change can indicate an object in motion, through the Doppler Shift. A light curve of spectra + timing can provide information on occultation as well.

If you put together any pair of the Image/Brightness/Spectra/Timing elements, you get a new data item to track. Consider these combinations:

- position + timing = movement
- brightness + timing = variability

- color + timing = change in composition or temp
- position + brightness = magnitude
- position + color = what and where
- brightness + color = more composition info

The Eternal Fight: Resolution Versus Brightness

You can capture something bright, or you can subdivide it to get more information, but you can't capture all the information in all possible ways.

The modern conflict is data bandwidth versus data accumulation. We are always starving for photons, either because the source is dim, or our detector is small or inefficient, or our download bandwidth is limited—no matter what, you can't get as many photons as you want.

We call this being *photon limited*, literally without enough photons to see the object. Most astronomical observations by the Hubble Space Telescope, for example, are photon limited because the objects themselves are very distant and thus faint.

We also describe being *bandwidth limited*: not able to deliver enough data. The object may be quite visible, but (like a slow internet connection) you simply can't access it. The STEREO mission, which uses two satellites with 11 instruments to stare at the Sun, is an example where there are certainly enough photons available, but we lack enough communications bandwidth to transmit the multiple Terabytes needed to Earth.

Active Detectors

So far, we have focused on traditional passive detectors. A source emits or reflects light, which the sensor captures. Active detectors provide their own light (or other) source. For example, a bat's sonar works because the bat emits a high pitched sound, then listens to its reflected echo to determine its position. It's a pity sound doesn't travel in space, or we could try this on a satellite.

Instead, Radar and LIDAR send out a beam of light (radio waves for Radar, laser pulses for LIDAR) then capture the return *bounce* and analyze how long that light took and how it changed to determine what is there. You could use a light source as a *flash* to either illuminate a target, or as a probe for a detector on the other side that looks at how the light was changed.

Active detectors, however, typically consume far more power than a passive detector, not the least because they need a passive detector element plus the addition of the active source. For picosatellites, use of active detectors is very cutting edge. Fortunately, in terms of interface, bandwidth calculation, and all the other work in this book, the same sensor theory applies. You may be setting yourself up with a more complex problem by building an active detector, but the principles remain sound.

Tradeoffs

Let's look at image brightness. Do you want a bright image with low detail, or very fine detail but a dim image? A high image resolution or a brighter image? Color information or a brighter image? Quick variability or a brighter image?

For timing accuracy, we consider both how long it takes to capture data, and how long it takes to prepare for the next frame. These are called *integration time* and *cadence*, respectively.

Integration time, a.k.a. shutter speed
: How long you have to stare to acquire the image

Cadence
: How quickly you can take successive images (due to readout time, processing time, etc.).

Both lead to your limits in timing accuracy. A sample timing accuracy calculation is: if it takes 3 seconds to acquire an image, your best time resolution is +/− 3 sec. If you can only take one image every minute, your best time resolution is +/− 60 sec.

The effects of timing issues include *smearing*—the source moved or changed during the exposure—and *gaps*—the source moved or changed in between frames (while you weren't looking).

There's a direct analogy to handheld digital cameras. Depending on the light level, you may or may not be able to see your target. There's a lag in between frames. If you want to pursue sports photography, you need a fast shutter speed and lots of light. Looking at detector tradeoffs involves a key set of decisions you need to examine before you shop around for a sensor. The issues to discuss are whether it is imaging or takes a single value, how many colors it can see, how accurately it can discern different brightness levels, and how quickly and how often it takes data. We'll work on these four—Imaging, Spectral, Brightness, and Timing—but first, we need to examine the sizes and framing of the actual targets we wish to examine. For that, we'll discuss Field of View.

Imaging Detectors

Detectors are specified with a given Field of View, which is the angular size they view in units of degrees (or the smaller increments of degrees: there are 60 arcminutes in 1 degree, and 60 arcseconds in 1 arcminute). A wide, panoramic field of view might be 180 degrees; a narrow zoom-in might be a degree or less. This is identical to the *wide/zoom* button on any digital camera.

Mapping Coverage to Pixels

In Figure 4-1, we show a sample FOV of about 25 degrees. At close range, we can only see 1 of the 3 objects, but it fills the whole field of view, so we'll get a nice big image of it. At a longer range, we can see all 3 of the 3 objects, but each one only fills a third of the field of view, so their size in the image (and thus their resolution) will be smaller.

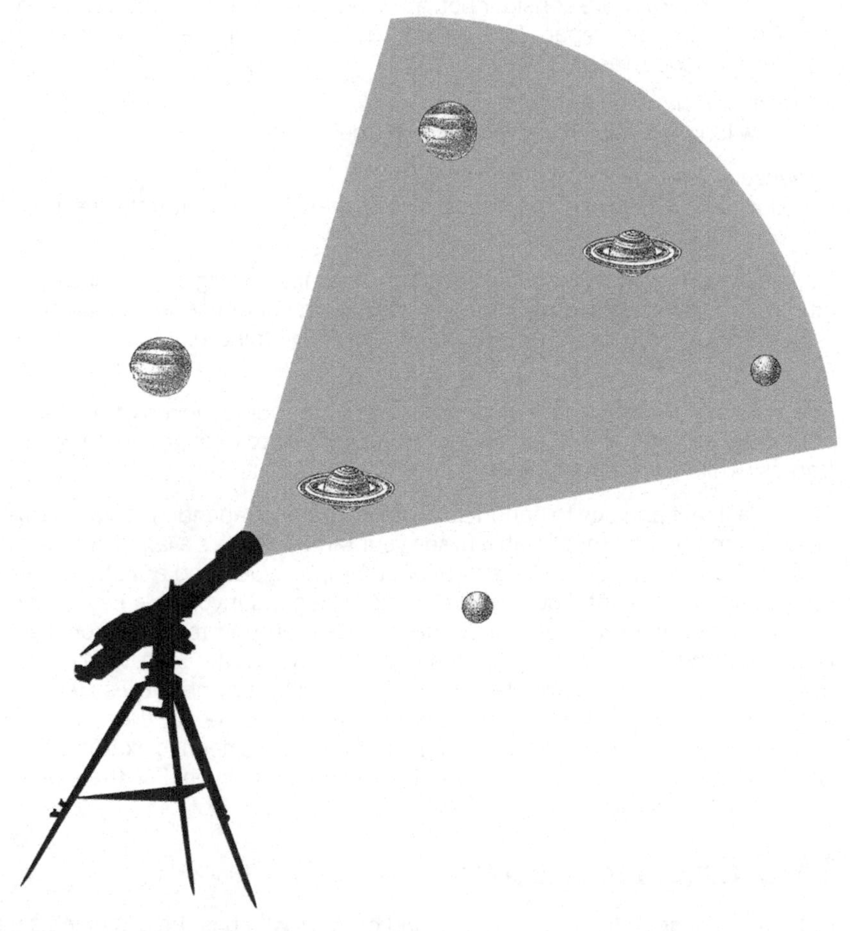

Figure 4-1. *Sample Field of View (FOV)*

For astrophysical missions, this angular field of view is all we need. For Earth observing and planetary and military sensing, however, we need to figure out how this angular measure maps to a Ground Field of View (GFOV)—how much area of the ground (in square kilometers or square miles) our detector will see.

To convert from angular FOV to ground FOV, you use simple trigonometry. If you are looking straight down, your FOV is basically a large isosceles triangle, where the ground is the base and your angular FOV is your top triangle angle. Using triangle math, the result is that the GFOV * 2 * (height_above_ground) = tan(angular_FOV/2).

For Earth Observing (EOS) maps, then, we can talk about our FOV in terms of area, of rectangular shapes. The calculation for "my image, map, scene that I'm capturing" to "detector elements" uses just multiplication to figure out our units:

```
Physical area/number of pixels = pixel size
```

or, assuming some physical area in units of kilometers (km):

```
Length_km/x_pixels by Width_km/y_pixels = (# of x pixels) by (# of y
pixels)
```

Phrased another way, an `A x B km` physical area mapped onto an `i x j` detector means each pixel covers `(A/i) km x (B/j) km`, as shown in these mappings:

```
Case 1: Square area, square detector
    Covering an area 100 x 100 km using:
        a 10 x 10 pixel array:    1 pixel = 10 km x 10km
        a 100 x 100 pixel array:  1 pixel =   1km x   1km

Case 2: Rectangular area, square detector
    Covering an area 20 x 100 km using:
        a 10 x 10 pixel array:    1 pixel =  2 km    x 10km
        a 100 x 100 pixel array:  1 pixel =  0.2km  x  1km

Case 3: Square area, Rectangular detector
    Covering an area 100 x 100 km using:
        a 20 x 10 pixel array:    1 pixel = 50 km  x 10km
        a 200 x 100 pixel array:  1 pixel = 0.5 km x   1km
```

Bear in mind we are assuming a straight downward (nadir) pointing. If you tilt your detector—as with the image of TRMM shown in Figure 4-2—your footprint calculation becomes more complex. However, the basic math does not change: deduce the geometry, figure out your footprint area, and divide by the number of pixels available.

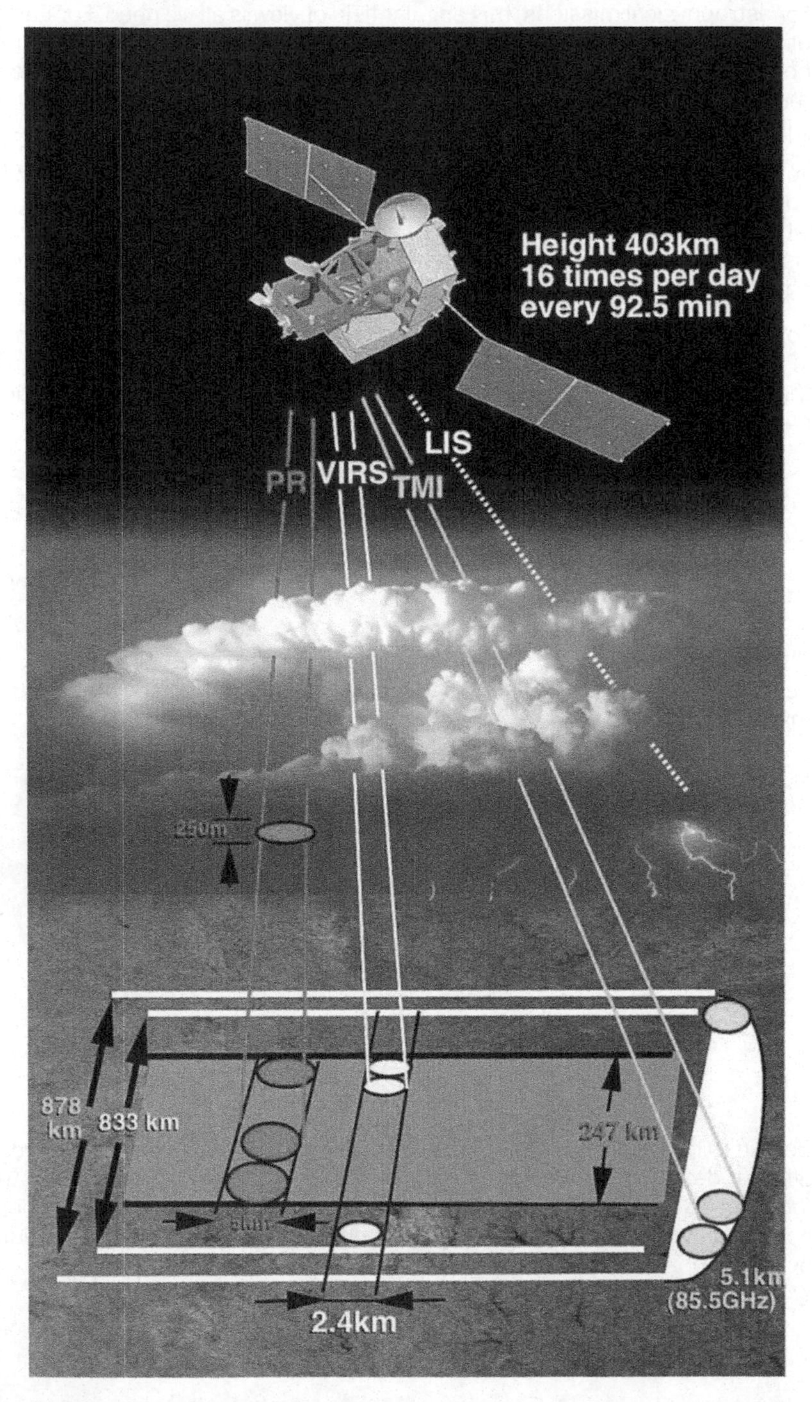

Figure 4-2. *Both nadir and off-angle pointings are assembled into 2-D image swaths as the satellite moves in orbit. Image courtesy NASA/TRMM.*

Sampling and Bandwidth Calculations

Most picosatellites are going to be bandwidth-limited rather than photon-limited. Photon-limited means you are looking at a faint source, such as astronomical objects. Bandwidth-limited, in contrast, means your sensors can get more data than you can transmit to the ground.

There are four factors that affect how much bandwidth you use:

Imaging (I): Are you taking an image (A by B pixels, requiring A*B data points), or are you just taking a single value like a temperature, an electric field strength, a magnetic field strength, or other single data point? Landsat-7's ETM camera is actually a scanner that takes a narrow 1-D line *swath* view and then (like a photocopier) combines subsequent lines into 2-D images. Typical Landsat-7 "scenes" might be 238×243 pixels. High definition video often uses 1,280×720 pixel images or larger.

(Landsat-7 covers the entire Earth with 57,784 scenes every 16 days: each scene is 233×248 pixels. That works out to roughly one scene taken each second. Because it takes scans and ends up covering the full Earth, you can really chop up its data any way you wish.)

Once we have a chosen detector size (our imaging resolution), we can figure out the bandwidth needed for our given quartet of resolutions (imaging, photometric, spectral, and timing).

Spectral or Colors (S): The number of colors or spectral bands you capture will multiply your bandwidth needs. If you take a single *band*, akin to monochrome images, there is no multiplier. However, if you wish to sample several *bands* (also called colors or channels), that will cost you. A color image, for example, requires three times the storage of a black and white (B&W) image: one data point each for the red, green, and blue elements that make up the full color.

If you are sensing past optical, you can sample discrete *bands* by providing filters that let you chop up a monochromatic range (such as *IR*) into bands. For example, the Landsat-7 optical plus infrared camera splits up the range of *IR* into 7 bands that include not just our ordinary 3 bands of red/blue/green color, but the near-IR (0.76–0.90 um), mid-IR (1.55–1.75 um and 2.08–2.35 um), and thermal IR (10.40–12.50 um). This gives more information, but also means they must transmit seven times the amount of data relative to just sending a B&W image.

Brightness levels (B): The accuracy of your detector in terms of number of bits of data used to map brightness sets your basic data unit.

If you need to store a number between 0 and 255, you have 256 distinct values. 2^8 is 256, so you can use that calculation to determine that you need 8 bits to store a value between 0 and 255. Similarly, if you need to store a value between 0 and 1,023, you'll need 10 bits of storage because 2^{10} is 1,024.

These numbers correspond to how well you can discern the contrast in a scene or element. If you are providing a coarse 8 different levels of brightness (equivalent to 8 different crayons to cover from *all white* to *all black*), you are using 2^3 data values, or 3 bits. If you want finer shading, say 64 levels, that is 2^6 data values and thus 7 bits. Landsat-7 uses 8-bit data (0–256) for its images. High-def video may be 12-bit data or higher.

Here are some common bit and byte measurements:

Data Size	Abbreviation
8 bits	1 byte
10^3 bytes	1 kilobyte
10^6 bytes	1 megabyte
10^9 bytes	1 gigabyte

Timing (T): One key element of a detector is how often you wish to sample it get to get data. If you take one data sample (one frame, one temperature measurement, etc.) per day, that's probably a safe, low bandwidth situation. If you instead need to send down high-definition video at 32 frames per second, that's going to be high bandwidth.

It can be weird to think of, for example, a thermometer as a *1-pixel detector*. You could rename *Imaging* to *Elements* if you wish. A thermometer is a 1-element detector; a 128–128 camera is a 128*128 element detector.

Our bandwidth calculation is pretty easy, as you just multiply these items to get a total, often converting from native *bits* into *bytes* (1 byte = 8 bits) and kilobytes (10^3 bytes), megabytes (10^6 bytes), gigabytes (10^9 bytes) or terabytes (10^{12} bytes).

You can determine the bandwidth your satellite needs with this equation:

```
Bandwidth = Image_size * Number_of_colors * Bits_per_pixel * Data_rate
(* 1 byte/8 bits)
```

Or, put another way:

```
Bandwidth = Imaging Res * Spectral Res * Photometric Res * Timing Res
(* 1 byte/8 bits)
```

So let's look at three detectors of wildly varying type.

- Temperature sensor, measures once/minute, uses 8 bits to encode number:

  ```
  Bandwidth = 1 temperature * 1 band * 8 bits * 1/minute = 8 bits per
  minute, or 1 byte per minute.
  ```

- Landsat-7 233×248 scene, 8 bands, 8-bit data, one taken per second:

```
Bandwidth = 233 * 248 * 8 * 8 bits * 60/min = 27.7 Megabytes per minute.
```

- High-def streaming video at 1280×720, color (3 bands), 12-bit data, 32 frames per second:

```
Bandwidth = 1,280 * 720 * 3 * 12 * (32*60) = 8 Gigabytes per minute!
```

Here's another sample scenario, a very plausible one you might fly. Assume a CubeSat with low resolution B&W images once per hour:

Measuring	**Resolution**
Imaging	32×32 pixels
Spectral	Grayscale → 1 channel
Brightness	256 brightness levels (8 bits/pixel)
Timing	1 frame each hour = 24/day

For this low-resolution CubeSat case, our data size = `32 * 32 * 1 * 8 * 24 * 1 byte/8bits = 24,576 bytes = 24 KB` per day!

If we downloaded this at the typical packet radio speed of 9,600 bps (a.k.a. 1,200 bytes/sec, a.k.a. 1.2 KB/sec), we need only a 17 second downlink.

Here's another sample, an EOS satellite with medium-high resolution color images every minute:

Measuring	**Resolution**
Imaging	1024×1024 pixels
Spectral	color RGB → 3 channels
Brightness	1024 brightness levels (10 bits/pixel)
Timing	1 frame each minute = 1440/day

For this medium resolution case, we require more bandwidth as we are gathering more data. Our data size = `1,024 * 1,024 * 3 * 10 * 1,440 * 1 byte/8bits = 5.6×109 bytes =~ 6 GB` per day!

You can see why picosats are often used for *in situ* measurements of the local space environment: temperature, fields, brightness, images, and fast data rates will balloon your bandwidth needs.

Fully understanding a sensor involves all four aspects—Imaging, Brightness levels, Spectral levels, and your timing. Each has synonyms: imaging can also be *number of detector elements*, particularly for non-picture sensors. Brightness can be called photometry, and photometric brightness is a synonym for "how bright light is." In contrast, you can also talk about *temperature brightness* or just about any variable and how strong its signal is. So you can think of brightness as *signal strength and ability to sense contrast*. A bit wordy, but it'll do.

And, of course, we talk about Spectral or color choices, which can also be called *bands* or *channels*—particularly if you are going past visible light to the longer wavelengths of IR, submillimeter, microwave, and radio; or the shorter wavelengths of UV, X-ray, or gamma ray.

This synonym soup of multiple terms arises primarily because different scientists and engineers working in different fields all have their own particular form of jargon. Only in the past few decades has there been more combined multi-instrument work, resulting in the need to learn many synonyms.

Timing is often timing, timing resolution, capture speed, shutter speed, cadence (time between observations), revisit time (time before you return to the same spot), and a myriad of other terms. In this case, each term has a very specific meaning, but they all work together to define *how quickly* and *how often* you measure your targets. Table 4-1 summarizes each category and what it lets you determine about your target.

Table 4-1. *Components and meaning*

Domain	Measures	Determines	Utility
Imaging	Spatial	Shape & size	Verify existence?
Brightness	Brightness	Temperature, concentration or density	How much is there?
Spectral	Color	Composition	What it is made of?
Timing	Timing	Dynamics	Rate it changes?

5/Detectors and Instruments and Sensors, Oh My!

The first question you'll get asked when you announce you're building a satellite is "What will it do?" While building the satellite platform is a fun engineering challenge, choosing and creating your satellite's payload involves both science and engineering. Payloads are fun, and what distinguish one satellite from another.

Some satellites will carry sensors that make measurements of either the area around the satellite, or something the satellite is pointing to. Other satellites will deploy a piece of gear that does something interesting or useful. A third category of satellite missions is there to explore technical and operational capabilities such as flying multiple coordinated satellites (constellations), testing engine designs, and other "no one has tried this" ideas (Figure 5-1).

Figure 5-1. *And sometimes, a single picture suffices. Hurricane Katrina, observed by TRMM. Image courtesy of NASA.*

We'll cover these three goals: Sensors, Deployable Payloads, and Operational Demos. Your satellite might have a bit of all three.

Some initial definitions are in order. A satellite has both a main platform or *bus* and a payload. The bus consists of the power system, an onboard processor, and a radio/antenna setup. Attached to this bus is your payload, the part that does something.

Assuming it survives launch and reached orbit, your satellite has a finite lifetime to complete your mission. This lifetime is determined by a) how long your orbit lasts, b) how long your satellite can survive being in space, and c) how long the power, bus, and payload last. Whatever its lifetime, you want your satellite to do specific, smart, intentional things. This book is about what your satellite does during its useful lifetime.

Attitude

Attitude refers to the ability to assess and control the directional facing of your satellite. It is distinct from being able to maneuver or reposition your satellite. Attitude for picosatellites is primarily a function of whether a) your satellite faces in a consistent direction and b) whether you can repoint your satellite.

Attitude is your satellite's *orientation*, while maneuvering affects your satellite's *position*. We do not discuss maneuvering in this book, but attitude can have a significant affect on your ability to capture sensor data.

There are two forms of attitude control: passive and active. Passive means your satellite has a specific facing in orbit without any control or operations required to obtain that. Active means that some sort of mechanical or electronic apparatus on the satellite allows you to set the satellites position.

Passive Attitude Control

Passive attitude control methods include:

- Free flyer (uncontrolled)
- Spinner
- Magnetically aligned

A free flyer is just that, flying free. It is ejected from the rocket and tumbles or coasts in its orbital groove for its entire mission. If your satellite does not have to point in any specific direction, this is acceptable. For example, a shiny silver ball of a satellite that you intend to bounce radio waves off of does not require a specific orientation.

A free flyer is often a cube or sphere. Its sensors require no specific pointing direction. My own "Project Calliope" is a free flyer, and we'll use light sensors to measure what its tumble rate and direction is.

More often, you may wish to have a known orientation of your satellite. A spinner is a satellite that is spun up by the rocket before being deployed. Conservation of angular momentum assures that the satellite will remain with its initial orientation as it orbits. A satellite deployed with a spin is akin to a gyroscope and retains its facing due to spin.

Magnetic alignment makes use of the fact that Low Earth Orbit (LEO) in particular has a magnetic field that goes from Earth pole to pole. For satellites put into a polar orbit, then, you can include magnets on the satellite to force the satellite to align with the Earth's magnetic field lines. This sort of satellite travels like a rifle bullet, its nose following the line of its orbit.

Because your satellite follows the track of its orbit with a specific forward-facing attitude, magnetic alignment is popular with Earth-observing picosatellites. The alignment assures you can have a nadir-pointing (downward-pointing) camera or other detector that always faces the Earth.

An astronomical mission that has passive pointing can operate in *survey mode*, where instead of pointing at a chosen object, it relies on the fact that its orbit is always changing to eventually map out an entire section of the sky.

Active Attitude Control

Active attitude control, in contrast to passive, enables you to point your satellite via ground commands or preprogrammed sequences to any particular pointing direction you wish. This can be useful for an astrophysical or solar mission, where you wish to target a specific object. It also enables any satellite to establish a known facing.

Many larger satellites with active attitude control use gyros: three or more rotating wheels that retain their orientation and can be sped up or slowed down to adjust its facing. You typically need three gyros to operate (one for each axis X/Y/Z or their Euler equivalents). Because mechanical parts fail in space, most satellites using gyros include spares. Due to the small available space in picosatellites, most picosatellites have not historically used gyros.

Magnetic torque bars are a method of *pushing* against the Earth's magnetic field to twist the satellite in a desired direction. They use the same principal as the passive magnetic alignment, except they can be altered mechanically or electronically to actively change the satellite's position as well.

Engines are what people often think of for positioning a satellite. While engines are good for maneuvers and changing orbit, they are not often used to change a satellite's facing. Changing attitude like this would require two burns: one to start a spin in the desired direction, then one to stop the spin.

During such a maneuver, you would risk getting exhaust on your satellite. There are workarounds for this, however.

Engines can include conventional rocket propellants that burn to provide the rocket effect. In theory, compressed air can be used as a rocket; however, pressurized containers are typically not flyable as NASA secondary payloads on NASA rockets and have not historically been used.

Pulsed-plasma and ion drives are a promising area of technology. They use a small amount of propellant (typically hydrogen gas) that is given a charge (ionized) then accelerated to a very high speed using an electric field. They provide a low thrust but can operate for a long time with very little fuel and high efficiency.

Solar sails and other experimental techniques are also methods of changing position.

For sensor work, we are assuming you do not have to maneuver the satellite, but may want to repoint it (change its attitude). This is, of course, so you can bring your sensors to bear on your target of interest.

Pointing Observations

One big issue to worry about if you are doing Earth-observing (or any) imaging is that your LEO satellite is likely moving at 7+ km/sec. If you need to stare at your target for any period of time, you will get significant motion blur. Even on Earth you can see this in the photos known as *star trails*. If you place a camera in a fixed spot pointing at the sky at night and take a long exposure, the stars will appear to move and create blurred trails.

This is the same reason that, if you look at a planet, star, or even the Moon in a telescope, you need to keep repointing your telescope or the object moves out of the field of view. The Earth is turning relative to the fixed background stars, so you have to counterrotate your scope to keep an astronomical item centered.

It's the same thing for a telescope in space, whether pointed out at stars or down at Earth. Your detector is moving, ergo you have to either shoot your picture very quickly, or be able to rotate to adjust for your position.

The excellent XKCD webcomic spinoff What If (*http://what-if.xkcd.com/32/*) runs the numbers for what the Hubble Space Telescope would need to get non-blurred images of the Earth. Basically, if you can rotate your telescope about 1 degree per second, you can keep the Earth centered in your field of view.

Note that *centered* does not necessarily mean *stable*. Accurate pointing requires that your satellite a) know its exact facing and b) be able to tweak its facing by small amounts to ensure everything is aligned. The problem you are facing with pointing is akin to trying to photograph a roadside object from a car while your driver is zooming around an oval race track. It's predictable, it's doable—but it's very hard. For picosatellites, it requires a fair amount of gear as well.

Contrast this with the existence of many polar weather satellites that take images even though the satellites are moving—and spinning. They solve motion blur in a very clever way, a method that is directly applicable to picosatellites. They use a single pixel detector and rely on the movement of the satellite as it spins to cover their entire target.

A single pixel imager is a very easy sensor to design and integrate, as it's basically a photometer or brightness detector. If your satellite has regular movement (orbit, spinning or both) AND you have very good information about your exact attitude (position) and pointing direction, then you can take each *1 pixel snapshot* and stitch them together to make an image.

Humans tend to think in pictures, and taking a camera like snapshot is certainly possible. Stitching together a series of 1-pixel data points is difficult given most picosatellites do not have strong knowledge of their attitude. Therefore, a compromise is to take a series of smaller images (perhaps 64×64 pixels, or 128×128 pixels) and stitch them together. Better yet, take a series of swaths or scans (such as 128×1 pixels) and add them up.

By stitching together smaller images, you have the advantage of not having to worry about motion or pointing, and can allow the satellite to passively point itself by orbiting and/or spinning. You put the data together on-ground. Small images are better than 1-pixel detectors because you can interpret where the images overlap more readily, since there will be overlap. This lets you juggle each image's position by matching edges, similar to solving a jigsaw puzzle.

Bear in mind that my own Project Calliope does no imaging, but instead reads the local space environment's magnetic field, the overall brightness level facing one side of the (tumbling) satellite, and tells us the temperature at the satellite as well. All of these are single data points sampled at a slow cadence, resulting in a satellite acting like an environmental space buoy. But if you must have images, know that a) it is doable and b) other satellites have done it!

Sensors

Sensors measure things. Some measurements may be about where you are, or *in situ* measurements. Your position, orientation, velocity, and acceleration are measurable. Your temperature, the brightness and electric and magnetic field around you, and the presence of any materials are environmental *in situ* measurements you or your satellite can measure.

You can also measure something that your satellite is pointing to, in a variety of wavelengths or in assessing particle counts or composition. Some of your measurements may require you bounce a known source of light (including radio) and measure the return signal to deduce what you're seeing (see Figure 5-2). All of these are ways of sensing.

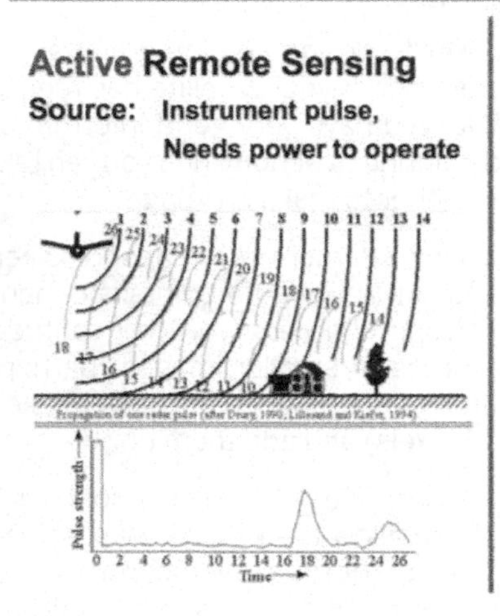

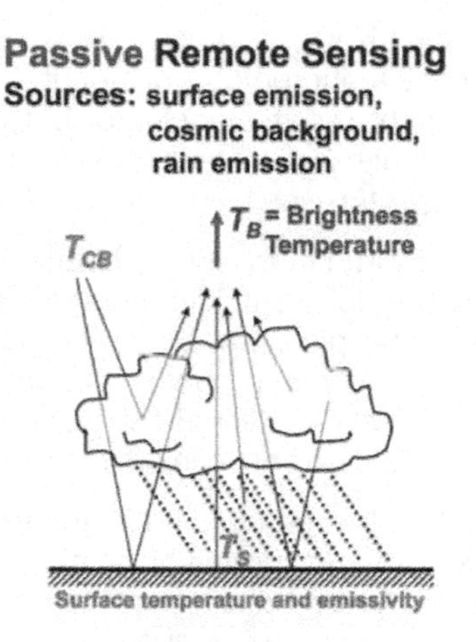

Figure 5-2. *Active (radar-like) sensors indicate distance, shape, and velocity; passive (telescope-like) sensors provide brightness and color. Image courtesy NASA/GPM.*

Sensing returns a data value. As a brief science recap, each single datum value has three components: a number, a unit, and an error range. A temperature might be 270 degrees Kelvin, plus or minus 1 degree. An Olympic race time might be listed as 9.63 seconds—but with an underlying accuracy of +/–0.001 seconds (and the millisecond portion of the number not reported). Even a photographic image consists of individual pixels that have a set of spatial and brightness units and error ranges attached.

When you choose and build your sensors, you will have to parameterize them in terms of the range and accuracy they measure, and capture that data with an understanding of the values and errors they contain. Since your sensors must talk to the CPU on your satellite bus, you will also deal with timing and sampling issues. All of this folds into the full data picture.

Sensors Versus Instruments

We will use *sensor* and *instrument* fairly interchangeably to refer to any device used to make a measurement. If you wish to be precise, a sensor is a device that returns a specific type of measurements as a series of individual data points. An instrument is a payload that is designed to return a specific set of scientific data. An instrument consists of one or more sensors to achieve its goals. Further, an instrument package or a telescope may include multiple instruments within it, each of which has one or more sensors. By way of example, the Hubble Space Telescope (HST) has multiple instruments that can be used with the HST's mirror—and one package, the initial Wide Field and Planetary Camera (WF/PC) instrument, included two cameras, each of which used four CCD imaging devices (as well as having 48 filters that could be selected). That's 8 sensors in 2 instruments making 1 instrument package on a telescope that itself has multiple instrument packages, but generally returning just 1 image at any given time. If you wish to use sensor and detector interchangeably, that's okay with us, in which case an instrument is a bundle of one or more sensors.

Deployable Payloads

A deployable payload is a non-sensor piece of equipment that does something other than making a measurement. Deployables can include transmitters, mechanical devices, reflectors, and even other satellites.

Transmitter

A satellite that simply orbits and transmits a message is the most basic fully functional satellite design. Without a transmitter or a way of returning data, a satellite is just orbiting debris.

The first artificial satellite, 1957's Sputnik, was a simple transmitter that emitted a steady 1-Watt radio beep-beep-beep at 0.3-second intervals (at 20 and 40MHz) for anyone on the ground to receive. At 58.5cm in diameter, it was not much bigger than a 3U CubeSat. Even today, a simple Sputnik-like transmitter is a popular Cubesat loadout because your basic satellite bus provides this capacity without you having to add any additional payload.

Modern concepts like microchip satellites and other CubeSat projects seek to send up multiple smaller transmitters that broadcast either preselected or user-uploaded signals (or Twitter-like tweet) on the amateur radio bands. Multiple smaller-than-picosatellite satellites are often called "Sprites," and are not per se limited to just being transmitters, but can be functioning satellites on their own.

Extending your picosatellite to be more than just a transmitter can lead to its use as a repeater or broadcast satellite.

Microchip Satellites

Microchip satellites are even smaller than picosatellites. A single CubeSat can contain many of these wafer-sized spacecraft. Here are three news articles reporting on chipsat work.

KickSat
: According to the *Cornell Sun* (*http://bit.ly/Z08IKd*), "A Sprite consists of solar cells, a radio transceiver and a microcontroller with memory and sensors. It has many of the same capabilities as a large spacecraft, just scaled down, but current versions are not as advanced as most other spacecrafts quite yet. *The first version can't do much more than transmit its name and a few bits of data—think of it as a shrunken down Sputnik*."

Sprite
: According to Cornell University's Mason (*http://bit.ly/Y9S7aZ*), "Peck's tiny satellites—nicknamed "Sprite" and measuring just one square inch —are mounted on the Materials International Space Station Experiment (MISSE-8) pallet, which will be attached to the space station in turn. The chip satellites contain sensors, a microchip, and an antenna to transmit collected data about the chemistry of the solar wind and associated radiation and particle impacts. In a few years, the MISSE-8 panel will be removed and returned to Earth, so Peck can see how well his little ChipSats fared."

Repeaters and Reflectors

The origin of amateur satellites started with the Orbiting Satellite Carrying Amateur Radio (OSCAR) series, which is still viable to this day. OSCAR set multiple firsts, including first amateur satellites, first satellite to be deployed as a secondary payload into a distinct orbit, and first satellite voice transponder. They are also the longest running amateur satellite series, with over 70 OSCAR missions and 5 currently active. Anyone doing CubeSat work is making use of work founded by OSCAR and their worldwide support organization, AMSAT.

OSCAR supports amateur HAM radio communication for both voice (FM and Single Side Band/SSB) and data (using packet radio and the AX.25 or APRS protocols). AMSAT's website notes, "Throughout the years OSCAR satellites have helped make breakthroughs in the science of satellite communications. A few advancements include the launch of the first satellite voice transponder (OSCAR 3) and the development of highly advanced digital *store-and-forward* messaging transponder techniques. To date, over 70 OSCARs have been launched with more to be launched in the future."

A *repeater* is a satellite that receives a radio signal and repeats (retransmits) it in a new direction, at a higher power, at a difference frequency, or all three. "Transceiving (and TDRSS)" on page 45 delves a bit into communications jargon (which we'll cover more in the forthcoming Book 4, on communications and ground stations). Functionally, we'll use "Repeater" as the term for any payload whose goal is to act as a radio relay for signals.

A repeater has three parts: an antenna for receiving; the repeater hardware that captures the signal and does any optional amplification, changing of frequency, or processing; and the transmitting antenna to send the signal out again.

The receiving and transmitting antennas can be the same unit or distinct units pointing in different directions, depending on your design needs. If your satellite is intended to receive ground signals and bounce them back to the ground, a single antenna suffices. If your intent is to take signals from a particular direction and repeat them in a new direction and thus act as a relay—perhaps taking ground signals and transmitting them to another satellite, or to a lunar mission—two antennas are needed.

Transceiving (and TDRSS)

A transceiver is hardware that is a receiver plus a transmitter. A transponder (literally *transmitter-responder*) typically is a transceiver that also may change the frequency of the signal and/or do filtering, and may handle multiple channels. A mode called *store and forward* allows for time-shifting a signal as well, receiving it and then transmitting at a later time.

The current pinnacle of *repeater* or *relay* satellites is the Tracking and Data Relay Satellite System (TDRSS) network, a set of space-to-ground highly available communications satellites that are quite out of CubeSat reach.

A reflector is a device that directly reflects photons—visible, radio, or other. Think *mirror* for optical light and a big shiny item with large surface area for radio reflection. Deploying a large mirror or reflective item creates a bigger and highly reflective picosatellite. Existing concepts have put out optical mirrors, radio reflectors, and giant inflatable mylar or metallic bags. The primary goal is to put up a reflective surface that users from the ground can use for interacting with your satellite on an individual basis.

Reflectors improve the chance of amateur satellite tracking and spotting, of being able to spot your satellite visually by eye or telescope. If optically reflective, a ground user can try to bounce a high-powered laser onto it to try and receive the beam and thus estimate distance (see also *lunar ranging* and *the FAA does not want you firing high-powered lasers into the sky without permission*). More commonly, amateur radio enthusiasts will bounce their radio signal off it to reach a part of the globe otherwise unreachable.

Solar Sail

While the larger NanoSail-D was the first successful solar sail deployment in space, already CubeSat-based solar sail testbeds are in the works. Surrey's CubeSail (*http://bit.ly/WnD6S9*) is a 3U CubeSat that is worth looking up due to its copious schematics on how it indents to store and deploy its sail.

Deploying a solar sail is a difficult task requiring very strong mechanical engineering skills. It's not as simple as inflating a reflecting balloon because it requires precise positioning and attitude control. You must take a large, thin, fragile material, unfold it accurately, be able to determine and establish your position and orientation, all while working with new materials. Given the difficult technical challenges of current solar sail technology, CubeSats are an ideal test platform for solar sails.

Operational Demo

Any test of a new attitude control system, an engine, solar cell, or a new architecture like chipsats, is by definition an operational demo as well. New satellite platforms and motive systems require new methods of operation. In addition to these avenues, there are other areas where operations—not payloads—are the primary concern.

One large area of interest is the ability to fly and/or coordinate multiple picosatellites. Called "Constellations," the ability to use multiple satellites is an area of active research. Since this is a sensor book, however, we will leave that topic for research purposes.

Are GPS Satellites Instruments?

To answer whether GPS satellites are instruments, ask, "Do they have sensors?" They don't. They are not instruments, as their intended goal is not to receive light for analysis. They just transmit. As a counter-argument, some have laser retro-reflectors to help with this. Ground stations use radio quasars as reference points.

If you're interested in learning more about GPS, there's a neat guide at this website (*http://www.beaglesoft.com/gpstechnology.htm*).

Sample Names

At a high level, sensor hardware can be broken out into dozens of categories. These tend to be fairly unique types, but in general can neatly fit into the Imaging/Spectral/Brightness/Timing schema. Here is a list of some such sensor hardware:

- Camera
- Calorimeter
- Proportional counter
- LIDAR
- Interferometer
- Spectrometer
- Gravity Wave Detector
- Magnetic sensor
- Electric field sensor
- Ion detector

When you get to the actual implementation of hardware, however, there are hundreds of different types, each unique, most overlapping. The Mars Science Laboratory (MSL) rover on Mars had multiple cameras, several detectors that were spectrographs, and even a chemistry lab. For many space missions, each instrument is a unique item that evolves from, but is not directly like, any previous detector. Some specific implementations of the above categories might be:

- CCD
- Silicon thermistor micro-calorimeter
- Gas scintillation proportional counter
- LIDAR
- Fabre-Perot plate interferometer
- Holographic Fabre-Perot spectrometer
- Tunneling magnetoresistance magnetic sensors

There are too many potential designs to list; it's easier to look at stuff that has been launched. Parameterizing a sensor using our chart can help cross-compare instruments to help you design your own sensor loadout:

Mission Domain
    EO, Solar, Astro, Planetary

Type
    Imaging, Spectrometric, Photometric, Timing

Pointing
    Pointing, survey, *in situ*

Wavelength regime (energy range, colors)
    Nm or mm or m, also ev, keV, or bands, or channels

FOV

Either an angular measure (in degrees or arc minutes or arc seconds) or a ground equivalence (in km×km). Extra-wide, wide, medium, narrow.

Resolutions

Spatial: low/med/high; spectral: low/med/high; photometric: low/med/high; timing: low/med/high

Description

For instance, "An optical camera using a 4×4 array of 64×64 pixel 8-bit CCDs, shooting 1 frame per hour"

6/Colors and Brightness

Most people, when you say *sensor* or *detector*, think "camera." A camera is actually an imaging detector (takes a 2-D or 3-D picture) that has some specific photometric accuracy (ability to capture different levels of brightness) as well as having a spectral resolution range—capturing a specific range of colors (either monochrome or full color).

Both color and brightness are very important. In Figure 6-1, we are looking at a 1,000km×1,000km area of Seldovia, Alaska using Landsat data at 30 meters per pixel resolution, so it is a very high-resolution image. In color with full brightness levels, you can clearly see the land, water, ice, and mountains.

Figure 6-1. *Seldovia, AK at high spatial resolution, high color resolution, high brightness resolution*

When we remove the color information (Figure 6-2), we likewise begin to lose our ability to interpret the data.

Figure 6-2. *Seldovia, AK, back to the high spatial and brightness resolution, this time turning off our color information. Image analysis becomes harder to interpret.*

So, when we drop color, we lose information. Much of Earth observing and of astronomy expands our color range past just visual as shown in Figure 6-3. In particular, infrared and ultraviolet light are very useful for examining different layers of material, which we'll cover more as we look at some of the utility of color.

Most color detectors actually are 3-channel detectors responding to red, green, and blue light separately. By combining all three, you get a full-color image. If you instead look at each color frame separately, you also have spectral information; cross-comparing individual colors gives you color ratios that can provide information on the target's composition.

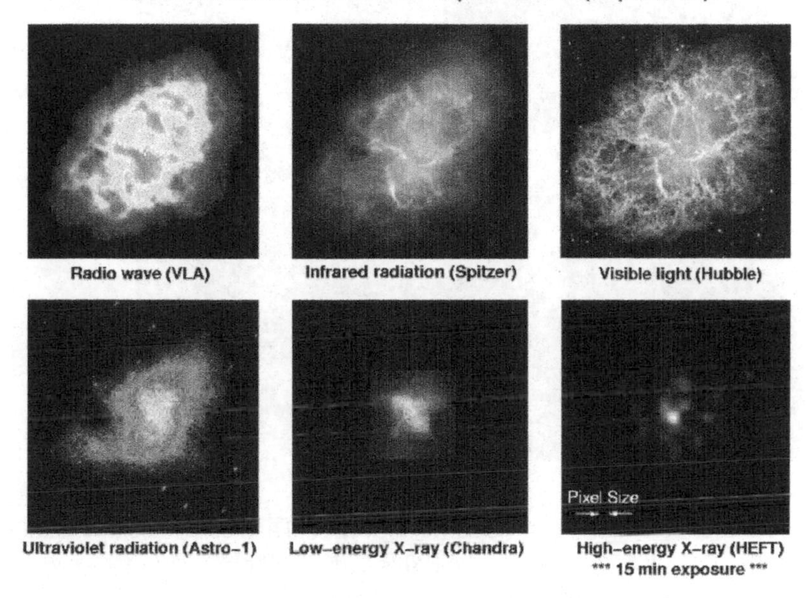

Figure 6-3. *Crab Nebula in various wavelengths. Image courtesy NASA.*

Brightness

The number of pixels and colors do not give us the full picture, however. We also need to look at how our range of brightness will affect our images. Going back to our Seldovia, AK, example, if we decrease our range of brightnesses (Figure 6-4), the picture becomes washed out, dull. We lose our ability to perceive differences in terrain because the contrast is lower. This is the effect of reducing Photometric (brightness) resolution.

Worst of all is poor photometric resolution combined with no color information (Figure 6-5). At that stage, you can no longer tell if you are looking at Earth or another planet.

Figure 6-4. *Seldovia, AK at the same high spatial and color resolutions, but with low brightness resolution. Our image is now muddy.*

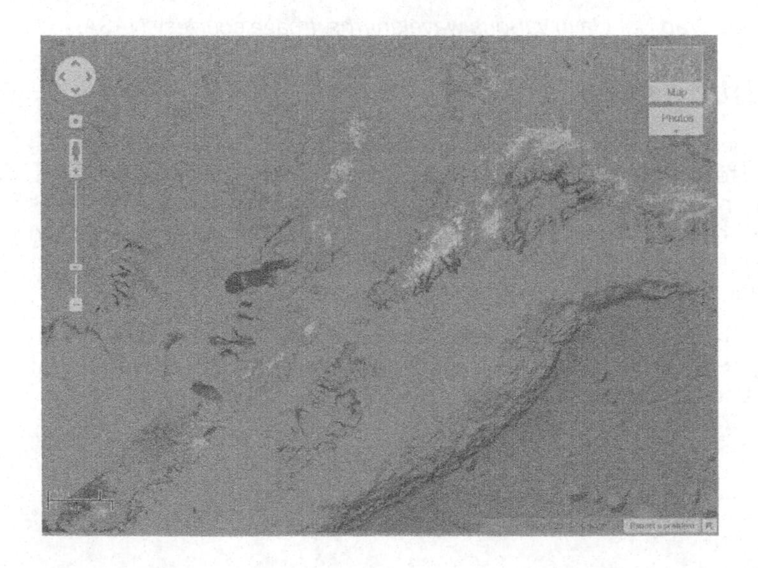

Figure 6-5. *Now we have poor color and brightness resolution. Although our image/spatial resolution is still excellent, the image is much less useful for analysis.*

We always want to allow for a wide range of brightnesses. Part of this is choosing appropriate upper and lower limits. The easiest example is measuring temperature with a thermometer. If you have a 6-inch thermometer and want to know today's weather, it is far more useful to have that thermometer scaled to go from 0°C (zero degrees Celsius, or water freezing) to perhaps 100°C (water boiling), than it would be to have it go from −273°C (absolute zero) to 1,000°C (way too hot). That larger range thermometer would be unreadable for most weather situations.

And, if you were using that thermometer to adjust your house temperatures, you might as well just have it go from 15°C to 25°C. Then you can easily read an exact temperature value, because the detector is tuned to the range of data you expect to take. Tweaking your upper and lower limits can let you compress your available resolution levels in a more useful fashion. When choosing or designing your sensor, make sure to understand all the parameters.

You can see this in practice whenever you use a voltmeter while building your satellite. Your voltmeter measures the voltage of a circuit. The meter is limited in its number of *brightness* levels—in this case, the number of digits it displays for your measurement. For an I2C electronics project, you probably will set it to a range of 0–10 volts, with 0.01-volt accuracy. However, use that same meter to measure your 120V house current, and you'll likely have a volt range of 1–1,000, with 1-volt accuracy. In both cases, your *brightness* range is broken into 1,000 subdivisions (either 0.01-volt increments, or 1-volt increments), as best suits the task at hand.

Note that both a thermometer and a voltmeter are *1-D* detectors; that is, they take no imaging information at all, just a single data value. You can consider them a *1-pixel* detector. Imaging, Brightness, and Spectral resolutions and ranges are all important in choosing your sensor.

7/Resolution, By the Numbers

Assume an ideal detector. Ideal = perfect 100% fidelity; thus it only returns data intrinsic to source with no noise, has perfectly square pixels, and there are no optical defects, *lens flare*, etc. That is, it's the sensor equivalent of physics *frictionless vacuum*.

Your target is a star with a possible exoplanet. Your detector receives 52 photons. We presume that's what your detector of a given size captures for the entire exposure. What happens when we try to catch it with different imaging detectors? Remember, no matter how we slice it, we only get 52 photons.

Detector Bins

Those 52 photons are incident on a single detector (say, 1cm×1cm). It is divided in this example into pixels. Each pixel is like a little bucket that collects the photons. We visualize the pixels shown in Figure 7-1.

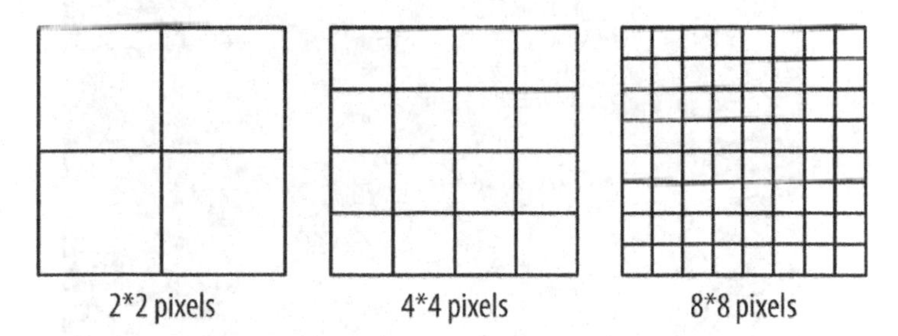

Figure 7-1. *Four detectors, each covering the same area, but with increasing spatial resolution*

Figure 7-2 shows a 2×2 pixel image. Figure 7-3 shows a 4×4 pixel image.

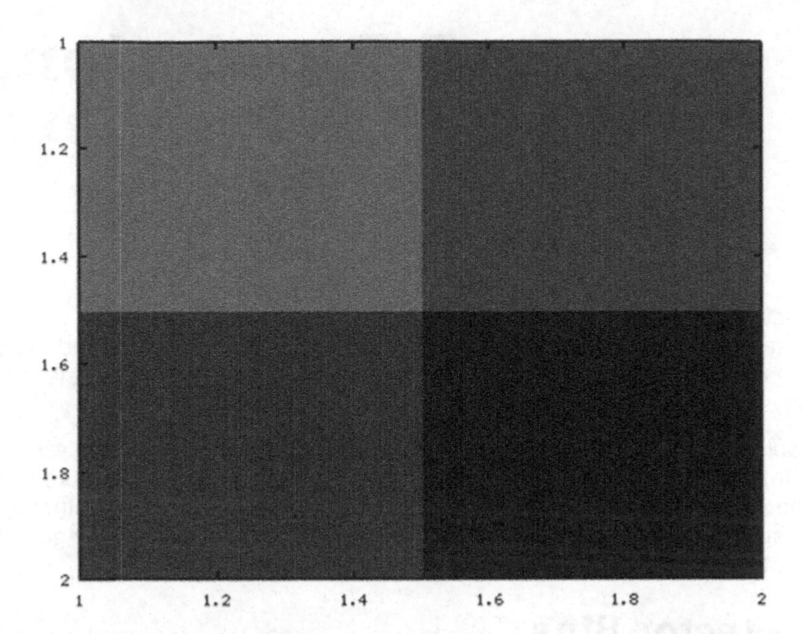

Figure 7-2. *2×2 pixel image*

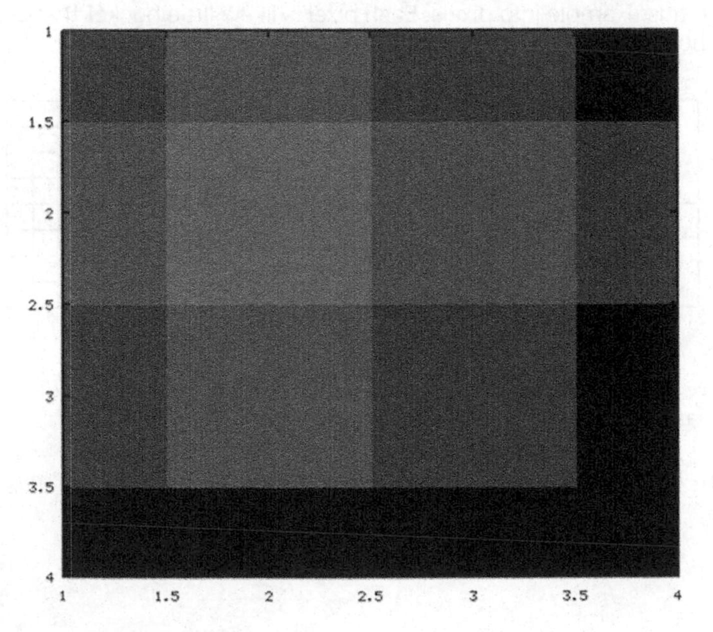

Figure 7-3. *4×4 pixel image*

Table 7-1 shows the number of photons in a 2×2 pixel image.

Table 7-1. *2×2 (photons)*

24	14
10	4

Table 7-2 shows the number of photons in a 4×4 pixel image.

Table 7-2. *4×4 (photons)*

4	6	4	0
6	8	6	4
4	6	4	0
0	0	0	0

Figure 7-4 shows an 8×8 pixel image. Table 7-3 shows the number of photons in a 8×8 pixel image.

Table 7-3. *8×8 (photons)*

0	1	1	1	1	0	0	0
1	2	2	2	2	1	0	0
1	2	2	2	2	1	0	2
1	2	2	2	2	1	0	2
1	2	2	2	2	1	0	0
0	1	1	1	1	0	0	0
0	0	0	0	0	0	0	0
0	0	0	0	0	0	0	0

Would you bet your academic career on saying this data shows an exoplanet?

What About Noise?

There are many kinds of noise in sensors:

1. Random (thermal) noise in the detector
2. Amplifier noise in the supporting electronics
3. Intrinsic statistical noise, where a given source emitting *x* has an actual emission spread of *fjx* (called *shot noise*, *poisson noise*, single counting statistics, etc. We'll cover this later.)

Assume a noise floor. We'll take an easy case. Let's assume our detector has a noise floor of 1 photon. That is, for any given cell in the detector, it will have plus or minus one photon due to detector noise, independent of the signal (+/− 1). We see that it changes our images—for the 2×2 noisy image (Figure 7-5), the source range = 4–24 and the detected signal range = 3–25.

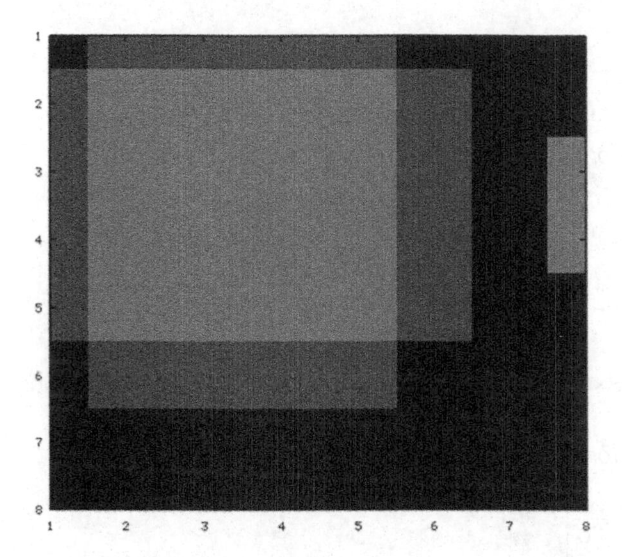

Figure 7-4. *8×8 pixel image*

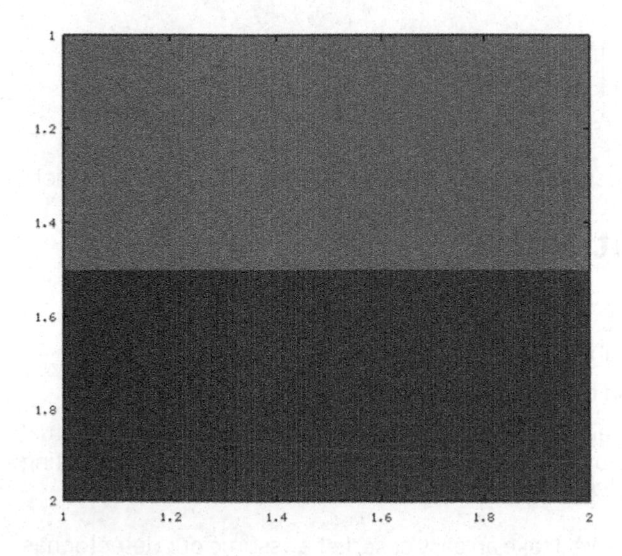

Figure 7-5. *2×2 noisy pixel image*

It's blocky, but there's something there.

For the 4×4 noisy image (Figure 7-6), the source range = 0 to 8 and the detected signal range = –1 to 9. We're getting marginal.

Now for the 8×8 noisy image (Figure 7-7), the source range = 0 to 2, which is so low that our detected signal range = –1 to 3, and we have essentially nothing useful in our data.

Now would you bet your academic career on saying this data shows an exoplanet?

At a Glance

There is no ideal detector (Figure 7-8). You have to decide what is important to measure, and what you are willing to give up to measure it. You are limited in detector size, detector configuration, bandwidth, and statistics. Also, it'll be slightly different for each wavelength, each instrument time, each object type you want.

Welcome to the puzzle of space-based sensors!

Filters

A filter is an object placed in front of a detector to block out some part of the incoming light (or signal). The two most common filters are spectral (or color) filters and image masks.

Spectral Filters

The most conventional usage of the term *filter* is usually taken to mean a filter that blocks some wavelengths (colors) of light so that you are only looking at the wavelengths you want. Such filters can serve to ensure a detector avoids contamination from non-desired light, or to optimize the spectral response.

A classic example is an X-ray CCD imaging detector. X-ray CCDs also respond to optical light, so if you observe a bright object that also emits X-rays, the detector will be so flooded with optical light that no X-ray data will be seen. Therefore, a thin aluminum filter is placed over the CCD. This completely blocks visual light, but allows incoming X-rays to pass through with almost no reduction. The CCD is now purely an X-ray detector, and serves its purpose without contamination.

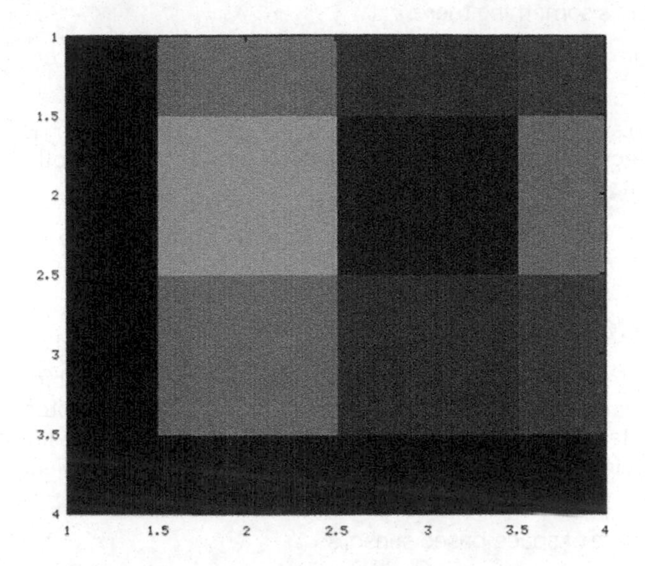

Figure 7-6. *4×4 noisy pixel image*

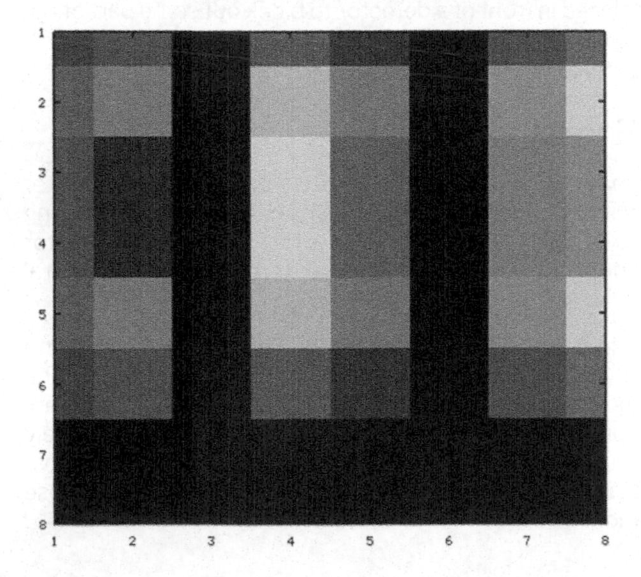

Figure 7-7. *8×8 noisy pixel image*

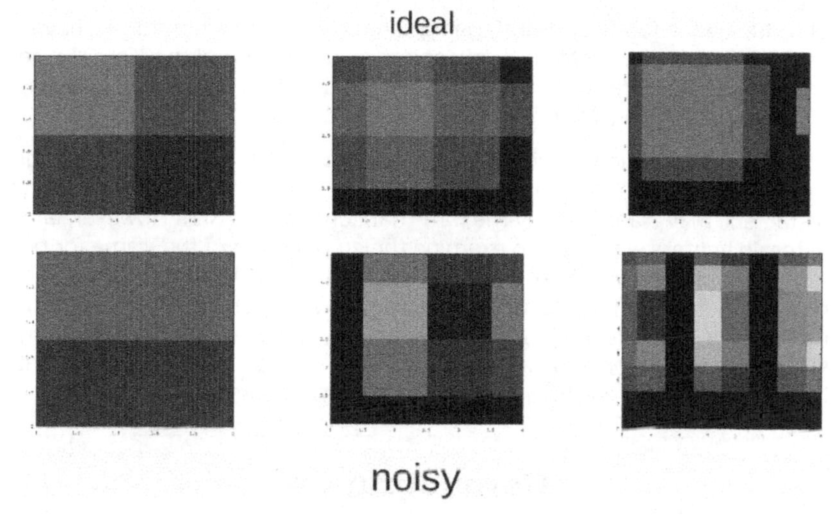

Figure 7-8. *At a glance, ideal versus noisy*

Filters can be added to any imaging or photometric detector to restrict the color range being seen. If you wish to (for example) see only green plant objects, you can use a filter that blocks all red and blue light. Your resulting data image now shows only plants.

The use of multiple filters is a technique often used to do imaging spectroscopy. By taking an ordinary full-range optical camera and interposing filters, you can take individual frames with color information.

A filter wheel is a mechanical device that will rotate any of a selection of filters in front of your detector. This allows one physical sensor to be used for different spectral channels, greatly increasing the potential scientific purposes. For picosatellites, I have not yet seen anyone attempt multi-spectral sensor works using a filter wheel, due to the additional size, power, commanding and mechanical needs this would require.

Image Masks

An image mask is a spatial filter. You block out some of our instrument's field of view (FOV). For example, you may be looking at a small target in a large FOV detector and do not want to waste bandwidth transmitting large empty areas. Therefore, you mask off the unused areas and only transmit the relevant parts.

Masking can be done digitally, only sending the part of the data that you wish. It can also be done physically—interposing a blocking material onto the detector's FOV.

Digital masking through only transmitting part of an image is an effective way of reducing the total bandwidth being sent down. By example, if you have a 1,024×1,024 pixel detector, but your observed target fills only a 100×100 pixel area, you can reduce your bandwidth needs to 1/100th the full frame by simply transmitting only the desired 100×100 pixel area.

The downside of such reduction is that you have made it so no one else can use the data for another purpose. For example, if an astronomer is observing a star, she may transmit only the part that contains the star. However, later asteroid hunters who want to examine the *empty* parts of the frame for potential asteroids can't do so, because the full frame information is lost. For research satellites, this reduces serendipitous discovery and potential data mining. For picosatellites, however, where we are very, very sharply constrained in resolution and data, such masking is rarely unfavorable and can be a very effective trick to optimizing your sensor usage.

Coronographs

One type of image mask used in solar and some planet-spotting work is a coronograph. These are used when there is a bright object (such as the Sun or a star) and you wish to observe a nearby faint object (such as the outer corona of the bright object, a nearby planet or asteroid, or a neighboring star).

You simply mask out the bright object so that the fainter objects can be seen. In practice, this means that (first) you are obscuring most of your field of view. Second, the bright object will have some degree of glow, vignetting, and general light contamination of your image because no blocking is perfect. But in practice, this masking technique is viable.

FOV masking can be used even when the detector itself is not taking an image; for example, with purely spectral or purely photometric detectors. Since all non *in situ* detectors have a FOV, all such detectors can be masked.

Digitization

Whether the sensor is analog and read by your CPU, or digital and doing its own on-board processing, the equation translating the real-world measurement values to the digital number (DN) stored and transmitted to you is straightforward:

```
DN = (Real_value - Lower_Limit) / Resolution
Real_value = (DN * Resolution) + Lower_Limit
```

`Lower_Limit` is the lowest value your detector can measure. `Upper_Limit` is the highest number it can measure. `Resolution` is how accurate you can make your brightness or signal strength measurement. `DN`, as mentioned, is the digital number returned by the computer. `Real_value` is the thing you are trying to measure. This is best explained by example.

Assume you want to measure house temperatures in the range of 15°C to 50°C with a half-degree accuracy, that is, you want to know within +/−0.5°C what the temperature is. Your LL = 15, UL = 50, resolution = 0.5 degrees. What sort of numbers do you get?

A temperature of 35°C would return `DN = (35-15)/0.5 = 40`. On the ground, you take your DN (40) and reverse it to get the temperature: `temp = (40 * 0.5) + 15 = 35 C`!

Now let's redesign this to measure the Sun, with a range of 0–1,000,000°C with accuracy to +/− 1,000°C. What does a measurement of 70,000°C give us?

```
DN = (70000 - 0)/1000 = 70
```

And, on ground, we reverse this:

```
temp = (70 * 1000) + 0 = 70000.
```

Done!

Just for fun, we can also figure out how many bits of data are needed to store your numbers. You may see one sensor returns *8-bit data*, another returns *12-bit data*, maybe a third has just *4-bit data*. This means you can have a number that ranges from 0 to $2^{\#}$ bits. 8-bit data returns from 0 to 2^8 = 0 to 256. 4-bit just provides 0 to 2^4 = 0 to 16.

The number of bits indicates how accurately you can differentiate between two values. Think of each number as the tick mark on a ruler. You can only read out exact integer numbers, ergo, your ruler is limited by the number of tick marks. A 1-meter ruler that has 10 evenly-spaced tickmarks can read only the values of 0.1m, 0.2m, ..., 1.0m, or down to 0.1m accuracy. The same ruler, if given 100 tick marks, reads off values of 0.01m, 0.02m, ... down to 0.01m accuracy. Those accuracies—the 0.1m accuracy or the better 0.01m accuracy—are the resolutions we discussed earlier.

Using our earlier Lower_Limit and Upper_Limit concepts, then, and Resolution, we can define the size of the data you need to store your desired work in bits:

```
2^n = (Upper_Limit - Lower_Limit) / Resolution.
```

So our thermometer for `Lower_Limit` = 15°C and `Upper_Limit` = 50°C with `Resolution` = 0.5°C; accuracy requires only (50–15)/0.5 = 50 numbers. The nearest power of 2 is 2^5 = 64 (covers 0–64). So we'll need a 5-bit detector.

For the second case, we have `Lower_Limit` = 0°C and `Upper_Limit` = 1,000,000°C with `Resolution` = 1,000°C, so we need the nearest 2^n = (1,000,000 – 0)/1,000 = 1,000, so 2^{10} gives us 0–1,024, and we'll need a 10-bit detector.

In practice, most detectors return a fixed number of bits and you can tweak the other numbers—lower and upper limits and resolution—to ensure that your detector captures the behavior you want in the range it is expected to have with the accuracy you desire.

8/Noise

"Signal" is the term for what you are trying to measure. "Noise" is extraneous information that can Noise by definition is considered random. If it's not random, but is a systemic element, you can often design around it.

Signal-to-Noise (S/N)

Instead of worrying about how much noise there is, we want to see how much there is relative to our signal. So we compute a Signal-to-Noise ratio (S/N). It is just S/N = Signal strength/Noise level.

You want an S/N ratio of at least 3, and ideally 10 or more, to have an effective detector.

If a single measurement or frame of data has too much noise, you can take multiple frames of the same source to reduce the noise. By adding together frames taken off an unchanging source, you reduce noise. This is because the signal (presumed constant) will always add to itself, while noise (presumed random) will have as much a chance of canceling itself out as adding.

This technique is why you always were taught to take lots of data for science fair projects in school, then average them. Finding the mean or average of the data is an effective way to reduce the inaccuracy. In this case, we call the Inaccuracy *noise* and want to maximize the signal, so we take more data.

Types of Noise

The base noise component of most photon-based and electric field signals has a statistical distribution that is called a Poisson distribution. Without going into statistics, or quantum mechanics (all fine topics), we can approximate a given signal as having four noise components:

1. Intrinsic Poisson noise, which goes as the square root of the signal
2. "Dark noise," or the signal you get from your detector even when you're not measuring something
3. Electronics and amplifier noise, which is often a function of design and temperature
4. Interference, or noise from outside sources

Let's tackle these in reverse order:

Interference from external sources

External sources (other instruments, your computer, radio transmissions, solar activity) can often be reduced with better shielding or better grounding. Some interference might only happen at certain times. For example, satellites in Low Earth Orbit going through the South Atlantic Anomaly will experience higher solar particle activity that will result in more noise. You can find more information on this in Book 2: Surviving Orbit the DIY Way (*http://oreil.ly/surviving_orbit_DIY*). Many satellites choose to not take data during these noisy passages.

Whatever interference you can't get rid of can be accepted (and will add to your *dark noise*, or noise your detector constantly makes).

Electronics and amplifier noise

You can reduce the noise electronics generate by improving your design. Using shorter wires and fewer circuits will result in less noise. Cooling most detectors and electronics will reduce their noise as well. Choice of parts, likewise, may result in less noise. And again, any noise you can't eliminate becomes part of your dark noise.

Dark noise

Outside of minimizing sources of noise as given in #4 and #3, you will still end up with some noise. Dark noise is inescapable, but at least you can measure it. Profile your detector by reading *data* while it's not actually measuring what you want to measure. If it's an imaging detector, take a picture with the lens cover on. If it's an *in situ* detector like a thermometer or radio field detector, take data when the potential real signal is at a minimum. This dark noise level will give you a good estimate of the minimum signal you can measure, because you will always have to measure something that rises above your noise.

You must get data above your noise level. If you have an electronic thermometer that has a dark noise (variability even when not measuring something hot) of +/− 0.5 degrees, for example, you are not going to have any useful measurements of temperatures below 1 degree. And you want to be at least measuring temperatures over 5 degrees—and measuring temperature differences that are at least 1.5 degrees apart.

Put another way, if your detector is only good to +/− 0.5 degrees, you cannot reliably distinguish between two temperatures less than a degree apart, and probably want to be sure they are farther apart. This is how noise, accuracy and error ranges interact.

Now, let's look at our intrinsic Poisson noise.

Poisson noise

Poisson noise (also called photon noise) is the statistical variation that will occur for even a steady source. It's an inherent part of measuring, because life is inexact. It goes as the square root of the signal. So, if you measure 9 times as much signal, you've improved your signal-to-noise ratio by 3! Check this math:

```
Photon Noise = sqrt(Signal).
```

Since S/N = Signal/Noise, then S/N = Signal/sqrt(Signal), therefore (with algebra), S/N = sqrt(Signal).

Ergo, if you go from a Signal of 1 (S/N = 1) to a Signal of 9 (S/N = 3), you've improved your S/N by 3. If you then increase your Signal by, say, 4, your S/N will improve by 2 to a S/N of 6—yes!

How do you improve a signal? You take more data, or you stare at your target longer to accumulate a larger reading.

Adding Noises

However, we still have that pesky dark noise to factor in. To add noises, you add them in quadrature, which is this:

```
N(total) = sqrt ( sum of all (noises squared))
```

Or for our case of just Photon Noise and Dark Noise,

```
N(total) = sqrt[  PhotonNoise^2 + DarkNoise^2 ]
```

The nice thing about this formula is it works with measurable values. You can measure your dark noise. Once you take data, you will have an idea of what your signal is. Ergo, you can estimate your accuracy.

So to recap, you want to measure signals that a) vary by more than 1 resolution element (from previous chapters) plus b) vary by more than your detector's inherent dark noise, plus you want to get enough signal that you have a reliable S/N ratio that brings confidence that your measurement is really what is happening, and not just a random flicker in the signal.

SNR Diagrams

Many detector datasheets will provide noise curves that indicate the baseline dark noise of that detector, and often how their detector responds to temperature. Detector efficiency also comes into play. CCDs in particular are defined by their Quantum Efficiency (QE), which is a percentage of signal they will measure, relative to the signal hitting them. In general, most detectors are dominated by dark noise for weak signals, then dominated by photon noise at higher signal levels.

One last consideration is that your detector is not going to be perfect at all wavelengths. We'll consider how it responds to different light—its response curve—when we discuss calibration.

9/Calibration

It's easy to calibrate, right? Here's how to calibrate simply:

1. Measure something you know well on the ground.
2. Set your detector limits so the image matches reality.
3. Now, anything new you measure is calibrated!

Unfortunately, with real hardware and looking at our fuzzy reality with imperfect information and the fact that photons are distributed statistically rather than deterministically, we run into problems:

1. A detector has a response curve. You need to calibrate along its entire wavelength.
2. You need to include any additional factors such as refraction, air, etc. (calibration often done in vacuum).
3. The instrument and/or optics may (will!) change a bit after the rigors of launch, deployment, and exposure to space.

Response Function

Each instrument will not mirror reality on a 1:1 basis. Instead, a sensor may be more sensitive to some wavelengths and less sensitive to others. The way a detector responds to different wavelengths is logically called its *response function*. This shows what fraction of received light gets *lost* by the detector. A perfect response is a value of *1*, meaning each source photon is accurately captured by the detector. A number less than 1 means the detector just isn't as efficient at gathering that particular wavelength of light.

For example, a sensor that has a poor response (0.5) to *blue* and a good response (1.0) to *red* will result in images that seem too red. All the red was captured, but much of the blue light was lost and is not seen. If you didn't know about its response function, you might think the actual observed target was red. If you properly correct for the response function, though (by artificially boosting any *blue* signal by a factor of 2), you can recover what the likely *true image* should be.

Technically, your response is a combination of every component the photons hit between when they are emitted and when they hit the detector. In a real-world sensor, this can mean the intermediary optics or mirror, any filters, any protective coverings on your detector, and the actual detector element itself.

A sample response function (Figure 9-1) is provided for a real instrument set. We show the response function for the entire signal chain on the Kepler planet-hunting mission. The key result is the purple *all elements combined* curve.

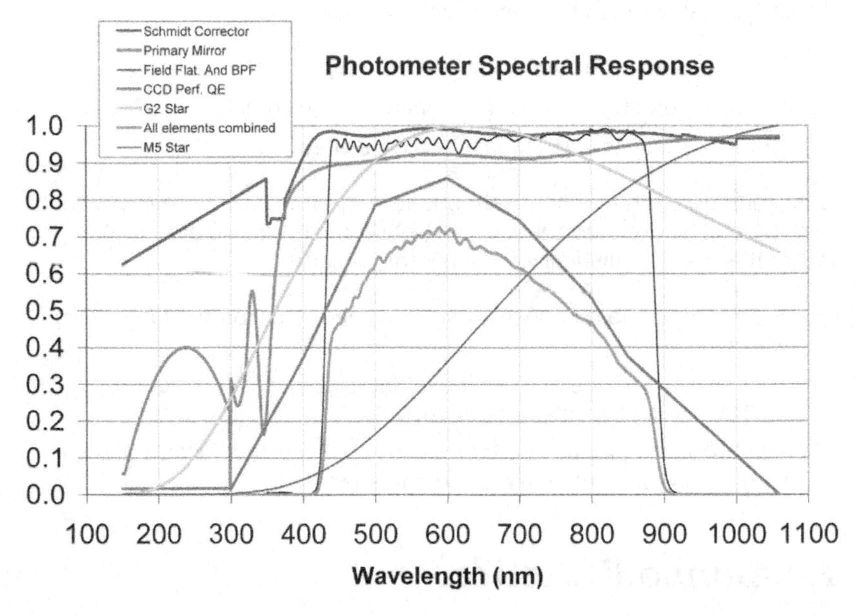

Figure 9-1. *Response function curves for the Kepler mission. Image courtesy of NASA.*

For Kepler photometry, this means that wavelengths shorter than about 420nm are simply not seen by the detector. Likewise, wavelengths longer than 900nm are not seen. Therefore, the instruments' effective wavelength range is 420–900nm.

Within the usable range of about 420–550nm, the response is uneven and low. So the detector will receive those photons poorly. From about 550nm–700nm, the detector has a nearly uniform response with about 60–70% efficiency. Finally, the 700nm–900nm range has the detector again getting worse at detecting photons.

By knowing this response function, you can take your data and attempt to recreate what the likely true source image is. The process of *deconvolving* the data using a response function is a data analysis task that is easily handled by computers, and is an essential part of scientific data processing.

For picosatellites, the steps to handling these facts are: first, to characterize your detector. This means obtain the response function, both through examining the technical data sheets on the detector and through actual ground measurements. Second, you want to filter or limit the light coming into your detector to block out undesired wavelengths. Finally, be sure to remember your detector's response function when you analyze your data.

Ground Calibration

For Project Calliope, I had to build a model ionosphere and replicate a spinning satellite to test what sort of data (music) to expect. In practice, this means rigging up a magnet to replicate ionospheric magnetic fields, putting a strobe for mimicking the Sun as seen by a rapidly spinning satellite, then seeing what sort of data my detectors spit out.

All instrument design requires you predict what range and variability your data has, then ensure the sensors you fly can cover that range. Sonification (converting that to music) has the added step of ensuring that the data provides some form and structure that is pleasing.

Calibration Protocols

Our situation is that we do not necessarily know whether what our detector outputs really are what the true object is. So we need some sort of *standard candle* that we can measure. We compare our data on the *standard candle* with what we know the actual true emission to be, then figure out what response curve our detector needs to reproduce that original source.

Solution!

Just pick a known target in the sky and use that as a *standard candle* to calibrate against! As long as your new detector returns the same data values that previous detectors have, you are calibrated!

Problem

If each detector is different, they will each see the *standard candle* slightly differently. If the new detector covers a different waveband, it will see it differently. If the new detector is *better*, how can you accurately tell how much better by comparing against an older detector?

Each wavelength sees something different. If the entire point of multiwavelength data collection is to capture different aspects of a target, how can you use the data of one instrument to cross-compare with another?

Standard Candles Aren't

Most potential *standard candles* vary. The Sun varies daily. The crab nebula varies. Vega varies. New York city lights vary.

If you have an improved detector, how can you tell if the old calibration is relevant? If this is your old calibration image, how can you tell if your new detector is more accurate, or just giving you a bad image in a different way?

So you do your best with the following:

- Do as much ground calibration as you can to map out the detector's response to known sources.
- In orbit, point to a previously observed source and try to reconstruct the previous observation using known information on how your instruments may differ.
- Try to reconcile any differences with an explanation: detector issue, optics issue, or intrinsic source variability?
- If possible, include a calibration source *on your detector*, either always in the FOV or something you can swing into the FOV. Recognize this means you lose some observing area or time due to calibration.

Remember Degradation

Instruments degrade or change over time. Ergo, you must periodically recalibrate.

In some sense, this is easier—just recalibrate so that your new data matches the earlier data taken of the same (presumed non-variable) object.

Real Versus "Book" Voltages

On paper, your power bus likely supplies a set voltage, your sensors require a specific voltage, and everything just works. However, in space, your parts may be operating outside of their usual temperature range, your battery charge/discharge cycles might result in a decreasing drift from their baseline voltage, and your voltages may differ from what you expected.

If you are using digital sensors that read (for example) pulses that are either 0V or 3.5V, a small different or drift in voltage will not cause trouble. However, if you are reading an analog sensor directly into your CPU and interpreting that sensor's voltage into a digital number, tracking voltage behavior of your main power bus is going to be crucial to return accurate numbers.

10/Protocols

A sensor just returns a signal, typically a voltage level or a digital series of pulses that encode a data number. If the latter, your sensor may be analog, returning a voltage that must be converted by your satellite's CPU. Or it may be digital, requiring you to listen at the appropriate clock rate to the correct input line in order to query and catch the signal.

Different sensor protocols—I2C (sometimes called 2-wire), TTL (serial pins), SPI, CAN, UART/serial/RS232 (asynchronous, requiring aligned clocks on both sides), and one-wire—all have different properties. In terms of readout speeds, the protocols, from fastest to slowest (in 8-bit bytes per second) are: SPI, I2C, parallel, serial, one-wire. In terms of complexity, programming, and syncing, they're all over the map.

Summed up, it's largely an evolution, of UART→SPI→I2C, so that's one reason I often suggest I2C. It's a *two-wire* protocol that makes it easy to hook up multiple sensors to your Arduino/BasicX/PIC. Alternatively, you can use TTL serial lines on the serial pins, but some Arduinos do not let you use both the serial and the ADC pins. Also, I2C, TLL, and serial are all designed to return different voltage ranges (–13V to 13V for serial, 0–5V for TTL and I2C).

Unfortunately, each protocol has several names, and the confusing habit of calling them by the number of signal wires persists.

Protocols by Number of Wires

SPI = four-wire serial bus (syn: SSI Synchronous Serial Interface)

I2C = two-wire serial

Three-wire = undefined, means several things

One-wire = like I2c, but lower speed and longer range; uses just data and ground, plus a cap to store the power charge while data line is active!

Bear in mind, you usually have to wire the sensors for ground, power, and clock, as well as the data pin(s).

Sensor Readout Theory

You then plug the sensor into the appropriate port, often a set of Analog-to-Digital (A/D) pins on the chosen CPU (Arduino, PIC, BasicX-24, etc). CPUs tend to have multiple A/D pins so you can put in multiple TTL or analog sensors, depending on the number of wires each requires. For I2C, there is usually a single pair of pins that can support up to 128 different I2C sensors, all chained together in parallel, yet operating separately. If you're dealing with an image sensor or have a PIC already doing some sensor interpretation, you may end up using the TTL/serial lines instead, along with code supplied by the manufacturer.

You then program the CPU to poll the appropriate sensor at the necessary rate and receive and interpret the voltages accordingly. Each I2C device has its own address, so when you poll, you poll a specific sensor, then get its information back. A good I2C technical tutorial can be found at this website (*http://bit.ly/12PY3ul*).

The main concern when choosing from the existing sensors is whether or not the sensor captures the data in the dynamic range that you need, with the resolution you need, at a readout frequency sufficient for your purposes. Also, it's easier to design and program if all your sensors use the same specs (i.e. all are TTL or all are I2C). Both TTL and I2C are well supported with Arduino and BasicX-24 boards, and are universal standards.

Wiring Sensors and Sampling

Most sensors have three or more wires. Two are power lines—a "power in" and a ground connection (to the entire satellite's ground). The next wire is the signal wire, of which there may be one or more. With more than one, you can measure a voltage across them. With one, you use the power ground as the baseline level. There may also be a clock wire that handles sensor sampling rates. But for starters, let's focus on the basic schema (Example 10-1):

Example 10-1. Sample sensor schematic

```
---power in----|            |=== voltage
               |--SENSOR -|
---ground------|
               |
---(clock)-----|
```

You read out the voltage from the sensor in order to get the signal. But how? At some point, your CPU has to poll (or ask for, then fetch the data from), the sensor. It may do this through a direct connection or via an intermediary *bus*. Let's examine each possibility.

With a sensor wired directly into your CPU board, you ask for data and you get it. It's a 1-to-1 relationship. When using a bus, you ask the bus to send you the data from a given named sensor. The bus queries the sensor, then gives you its data. It's a 1-to-many relationship.

With a direct connection, your sensor is outputting voltage at all times. It's always on (at least, while you're supplying power to it). Therefore, you can only read out data when you tell the CPU, "Hey, check what that signal voltage is."

Typically, your sensor will be hooked into one of the Analog-to-Digital ports on your CPU. This pair of pins does exactly what its name suggests: it reads an analog voltage delivered by your sensor and converts it to a digital number that your CPU can process.

Sensors that use a bus protocol, like I2C, differ from a pair of bare wires hooked straight into an ADC. Instead, you hook up multiple sensors to the chosen input pin pair. The advantage of using a bus is that you can hook up more sensors than you have direct connections. If (for example) your CPU board has eight ADC pin pairs, you can hook up, at most, eight sensors using a direct connection. Each sensor can be polled (read) by the CPU, and you handle what gets read through programming. You might instruct the CPU to "Read the sensor at pins 1–2," or "Read the sensor at pins 3–4 instead."

When using bus sensors, you daisy-chain all the sensors so they're wired into the same two pins, but you give each of them a different unique address. The sensors are smart enough to a) remember their own address and b) ignore any requests for data that aren't specifically sent to them. In order to collect data, you send a request to all the sensors saying, "Sensor with address [X], please give me your data," then you get only the data from that one sensor, while the others remain silent until you specifically ask them for data.

Clocks and Sampling Rates

You get data as often as you sample (or read out) the sensor.

Your sampling rate is how many times you receive a specific data value, per second. It is often expressed in Hertz (Hz), which is the reciprocal of time: frequency (in Hz) = 1 / time (in seconds)

So reading out twice per second means you sample it every 0.5 seconds, which would be a frequency of 1/(0.5), or 2 Hz. Reading out 4 times per second is 4 Hz (sampling every 0.25 seconds, so frequency = 1/0.25 sec). Reading out 60 times per second is 60 Hz. Reading out more slowly—say, once every 5 seconds—would be (1/5 Hz or 0.2 Hz).

The readout rate is determined by two factors—the clock speed of your CPU, and (potentially) the sampling rate of your sensor or sensor bus. The CPU clock speed is very important, because you cannot read out faster than the CPU can sample data. CPU clock speeds are often set by a timing crystal and are part of the CPU board's specifications.

For example, an Arduino has a clock speed of 16 MHz, while the BasicX-24 has a clock speed of about 8 MHz.

Synchronous and Asynchronous

In cases when you can poll a sensor at any time, you could say it is asynchronous, in that you do not have to worry about when you ask for data. Whenever you ask, it's ready.

Some sensors are not always "on," but will send data only when you ask them. In these cases, you have to tell your CPU to ask them for data, then wait for the response. This call-and-response is a form of asynchronous communication; it just adds the detail that you have to request sensor data before you get sensor data back.

Some sensors are always "on" but have their own clocks and send sensor data at their own speed. This adds the complication that you have to make sure you are listening when they are sending. This sort of communication is called synchronous, meaning you have to synchronize when you listen with when they send.

Now that we've covered theory, let's look at the most common sensor protocols, and how to wire them into your setup. Note most ADCs on your CPU are fine as TTL input/outputs.

I2C

I2C, sometimes called *2 wire*, is a bus protocol that lets you add many sensors onto one set of CPU board pins—as many as 128 (assuming you're using the standard 7-bit interface). You assign an address (name) to each sensor. When you want data, you ask for data from that specific sensor. Because it is called *2 wire*, it may surprise you that there are 4 wires involved, but as we discussed in "Sensor Readout Theory" on page 74, that's because each sensor will share a ground wire and (if needed) a power supply.

An I2C sensor sends its signal as a voltage pulse between 0 and 5V. Therefore, you'll need a stable 5V power supply, a good ground connection, and (ideally) an I2C port on your CPU hardware board. We also recommend you use an existing I2C library written for your CPU, as that neatly abstracts all the tech items into human-friendly *read* and *write* commands.

Your CPU board will have two pins, which you will use for all your I2C sensors. They are called SCL (*standard clock line*, or *clock*) and SDA (*standard data*, or *data*). You'll have to wire both SCL and SDA to your 5V power supply and include two *pullup* resistors (marked `Rpullup`) in order for the bus to work. (This is because the chips measure data as *negative voltage* or *voltage drop*, so you need to give it a reference voltage. Most importantly, just remember you have to supply this 5V reference voltage to both lines.)

Values for the two Rpullup resistors either a) depend on which reference document and which CPU board you are using or b) really don't matter much—just pick a value and then measure your output signal. Most people use 1,800 ohms, 4.7K ohms, or 10K ohms, according to Robot Electronics (*http://bit.ly/12PY3ul*). In general, try the smaller resistance and increase until it works—meaning your sensor can return a full value of 0–5V depending on what the sensor is reading.

All your sensors will be installed in parallel (Example 10-2):

Example 10-2. I2C bus

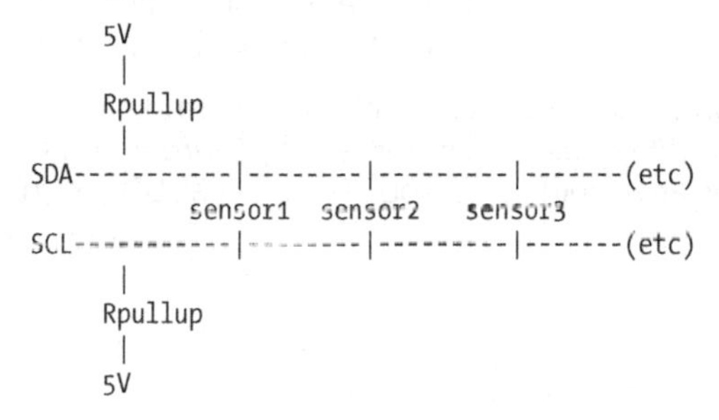

The shared ground wire (not pictured) is ground, or 0 volts. The shared power wire (also not pictured) is typically 5V DC. Try an Rpullup of 1,800 ohms.

Each sensor can actually have data written to it (such as assigning it a different address or setting internal configuration variables) and perform its usual task of having data read from it (giving you the measurement data). The general programming flow is the same for both reading and writing:

1. Send a *start sequence* pulse.
2. Send the address of the sensor you want.
3. Either send the data or read the data.
4. Send a *stop sequence* pulse.

Obviously things are a little more complex in practice. Fortunately, there are I2C libraries for the Arduino, for many PICs, and for BasicX-24 that abstract information so you can really perform the above commands without worrying about the electronics details. Since this book is intended for hobbyists, not electronic engineers, I highly recommend you use the libraries. One advantage of using an existing protocol like I2C with Commercial Off-the-Shelf (COTS) sensors that support I2C, is that you can focus on the design and integration and not worry about protocols.

Most sensors will come from the factory with a preassigned address. You can change that via programming, but for now we'll assume you will use it as preconfigured.

I2C has a standard clock speed (SCL) of up to 100KHz (100,000 times per second), and higher speed modes exist. Remember, that is how fast I2C can respond to queries. You are limited to the slower of your I2C or CPU speed—you can't read faster than the slowest part of your system. If you have a very complex sensor (with its own on-board chip doing extra processing, for example), you may be limited to a slower readout time. In particular, a slow sensor reading out lots of data may cause trouble if it cannot send its data within a single SCL clock pulse. The I2C solution is that slave sensors can artifically keep the SCL line *low* using a technique called *clock stretching*. As long as your control program is smart enough to deal with clock stretching, and does not send out new requests until the slave sensor indicates it is done (by releasing the low SCL line), you will not have synchronization problems.

So how do you tackle this with the libraries? Here's some sample routines using the Arduino I2C (*http://bit.ly/16sjgte*) and Wire (*http://arduino.cc/en/Reference/Wire*) libraries. On an Arduino Uno, you use pin A4 as SDA and pin A5 as SCL:

```
functions =
  begin()
  requestFrom()
  beginTransmission()
  endTransmission()
  write()
  available()
  read()
  onReceive()
  onRequest()
```

Here is a sample of the Arduino serial library (*http://bit.ly/13Wb6tF*), where all Arduino boards have at least one serial connection. Serials are often also called UART, and can use digital pins 0 (RX) and 1 (TX):

```
functions =
 available()
 begin()
 end()
 find()
 findUntil()
 parseFloat()
 parseInt()
 peek()
 print()
 println()
 read()
 readBytes()
 readBytesUntil()
 setTimeout()
 write()
  (with examples!)
```

To be complete, the BasicX-24 protensecode.zip library and better i2c-v2.bas contain I2C code so you can immediately use I2C with that processor. The functions are:

```
functions =
 I2cByteWrite()
 I2cread()
 I2cWordRead()
 I2cOutBye()
 I2cInByte()
 I2cStart()
 I2cStop()
```

I2C Sequence Primer

I recommend the Web for more formal I2C tutorials (search for *i2c tutorial* or *i2c bus specification*), but here is a summary of what actually occurs when using I2C. I2C is a *Master Slave* protocol, in which your CPU board is typically the master and each sensor is a slave. The CPU initiates a *start sequence* by toggling first the SDA low, then the SCL low. You then perform either a read or write operation. You wrap up with the *stop sequence*, which first sets the SCL high, then the SDA high.

Remember how we said sensor addresses were 7 bits and ran from 0-127? And now we're saying you send a byte (8 bits) to choose the address? Surprise. You send a leading *0* in the least signficiant bit (LSB) when starting either a read or write. The *0* means write but is, oddly enough, used for starting both a read and a write. When you are doing a read, you'll also send a *restart* in the middle with a leading *1* in the LSB, indicating that you intend to read. The nice thing about this is that all your messages (address, register, data) are always 1 byte (8 bits) in size.

To read, you perform the *start sequence* and send the 8-bit address (a *0* in the LSB, indicating *read*, plus the sensor's 7-bit address). You then send a byte indicating the address of a register you want to read. Now, instead of sending data (as you would with a write), which would screw things up, you send a *restart*: the 8-bit address, consisting of the sensor's 7-bit address, plus a *1* in the LSB, indicating *ready to read*. Now you can read as many data bytes as you wish. You close with the usual *stop sequence*.

The main reason to write to a sensor is to either set its address, or set internal variables that may define its upper and lower limits, its sensor resolution, or any other properties allowed by that specific sensor. There is no generic list of things you can set; each sensor implemented by the manufacturer is different and you'll need the spec sheet or manual to know what you can configure for that sensor.

To write, you initiate the *start sequence* as before, then send the 8-bit address (a *0* in the LSB, indicating *write*, plus the sensor's 7-bit address) to indicate which sensor should listen to you. You then send the information for the data register you want to write to, followed by the actual data (one or more bytes), and finally close with the *stop sequence*.

I2C Examples

Here is some sample I2C code for Arduino using the LiquidTWI2 lean high perm I2C LCD library for Arduino (*http://bit.ly/ZQcyXu*) (builds off LiquidTWI:)

```
#include <Wire.h>
#include <LiquidTWI2.h>
LiquidTWI2 lcd(0); // 0 = i2c address
void setup() {
  lcd.setMCPType(LTI_TYPE_MCP23008);
  lcd.begin(16,2);
  lcd.setBacklight(HIGH);
}
void loop() {
  lcd.print("Hello World");
  delay(500);
  lcd.clear();
  delay(500);
}
```

Here is the sensor spec for the I-CubeX Orient3D via a USB-microDig:

```
  sensor sampling interval: 100ms
  sensor input sampling resolution: 7 bit

Orient3D = 3 I2C lines as analog or all via 1 I2C bus
  Sensor output =
    Magnetic_field * sensor gain * cos(angle between field and sensor)
  returns 16 bit int (32767 to -32768)
  gain = 1/64, 1/32, ... 1/2, 1, 2
G = 31.24 Counts/uT at gain=1
combined,
  heading = arctan( Sx/Sy)
  pitch = arctan (Sx/Sz)
  roll = arctan (Sy/Sz)

(analog output: An = 5V * (sn + 512)/1024
```

TTL/UART/Serial/RS232

UART and TTL lines, serial lines, serial ports, RS-232 serial connectors, and DE-9 serial port connectors all follow the same schema. TTL and serial lines are similar, but have different pin voltages and pin orders.

A TTL line ranges from 0V to 5V, and is designed to output only a low (0–0.8V) or high (2–5V) signal, essentially producing digital (off or on) pulses. UART is usually 0–3.3V or 0–5V. You connect TTL point-to-point, typically connecting the *transmit* line (TXD) on device 1 to a *receive pin* (RXD) on device 2. It can be wired bidirectionally, since *transmit* and *receive* are independent circuits.

An RS-232 port has multiple pins, including TTL, ground, and power; of which some or all can be used for data. RS232 is asynchronous and requires that both the CPU and the sensor have clocks that are aligned—their clock speeds are multiples of each other. As with TTL, *transmit* and *receive* are independent circuits.

The RS-232 standard allows signals of either +3V to +15V or –3V to –15V. The formal spec allows for 25-pin connectors, though 9-pin serial connectors (DE-9) are common. Laptops and PCs are really not made with serial port connections anymore. Often, RS-232 is set as –12V to +12V (or –13V to +13V); it's all the same. Chips like a max232 or max3232 will convert TTL levels to RS-232.

Technically, both TTL and serial can operate synchronously or asynchronously. However, for synchronous operations, the devices share a clock line and regularly send just data or blanks. If you operate them asynchronously, you send start data and hope that the two devices are polling at roughly the same time. In asynchronous mode, the two devices must still have roughly coordinated clocks, else signals will be lost (see Examples 10-3 and 10-4):

Example 10-3. TTL spec

```
pin 1 gnd
+5 V
txd (transmit data)
rts (request to send)
rxd (receive data)
cts (clear to send)
+5 v
gnd
```

Example 10-4. DE-9 pinout spec

```
3 = txd
2 - rxd
4 = dtr data terminal ready
1 = dcd carrier detect
6 = dsr data set ready
9 = ri ring indicator
7 = rts
8 = cts
5 = common ground
```

If you are using a computer that doesn't have a serial port, you can find a USB-to-serial port converter, which will theoretically allow any computer's USB port to work as a serial port. However, a professor of electronic engineering has warned me that there are often problems with this, particularly in getting clock rates and synchronization to work.

SPI

The Serial Programming Interface (SPI) (*http://arduino.cc/en/Reference/SPI*), is a synchronous, serial Master-Slave protocol, which means a single master (typically the CPU) can handle multiple slave devices.

It uses 3 shared lines:

MISO
: Master in, slave out—used for slaves to send data to master

MOSI
: Master sends data to slaves

SCK
: Serial clock—the clock pulses to sync. Each slave has a SS (slave select) line that the master can use to enable or disable it. Therefore, though all the data is broadcast to all the slaves, only those slaves whose individual SS line is active will listen and obey.

The SPI library is standard with Arduino, and its bus is built into BasicX-24.

Controller Area Network

The Controller Area Network (CAN) spec is used in automotive work and is rumored to be making inroads to the DIY community. For Arduino purposes, the "canduino" library (*http://bit.ly/Yo6nMs*) in beta, is usable, with a custom Arduino board.

CAN is a multi-master protocol, which means each node can send or receive via broadcast. Therefore, each node listens for calls from any node that addresses it. CAN pins are as follows:

```
most common = using a 9-pin connector
  2 = CAN-Low (CAN-)
  3 = GND
  7 = CAN-High (CAN+)
  9 = CAM V+ (Power)
```

Musical Instrument Digital Interface

Just for fun, I'm including the Musical Instrument Digital Interface (MIDI). MIDI is a protocol used heavily in the music industry to enable synthesizers and input devices to communicate with each other. It uses a 5-pin controller (one-way) with only three essential lines—ground and a +5V pair—along with two optional lines for custom use. MIDI operates at 31.25 kHz, which was chosen because it's an exact division of 1 MHz.

There is a library at this site (*http://bit.ly/Z2pQIG*) for doing MIDI i/o via a serial port, and a tutorial at this site (*http://bit.ly/Y9ST7W*).

MIDI is a specific format for serial data at 31,250 bps (fixed), and uses a 10-bit message consisting of one start bit, eight data bits, and one stop bit. There are two types of data sent: channel and system. Because the protocol operates one-way, a device can only send information (via its MIDI out) to another device's MIDI in; therefore, you need two cables in order to work bidirectionally. However, any device can have up to 16 simultaneous separate data channels that serve as independent devices.

The channel data message indicates which of 16 different channels to listen for, and includes what we'd consider actual sensor data: note on (pitch+velocity), note off (pitch), polyphonic aftertouch, control change, mode change, program change, channel aftertouch, and pitchbend. The other message type, system, includes the mnemonics of System Common, System Real Time, and System Exclusive, and is broadcasted to all 16 channels.

Although the Arduino is well suited for MIDI, the BasicX-24 needs a replacement clock with an 8mhz clock chip in order to synchronize.

About Standards

The thing about standards is that it doesn't often matter what the standard is, it only matters that it exists. At a DIY level, you will typically be served best by picking a protocol that is 1) supported by your hardware and 2) supported by your sensor.

Consider this an extension of the "use it because you know it, because it's installed, and/or because it's best" criteria. The primary emphasis for your sensor should not be on which protocol is optimal; rather, the primary concern should be the sensor itself.

The sensor's fidelity, the amount of documentation available on the sensor, its temperature performance, its response curve, and its general hardware capability are the reason you choose a given sensor. Given that I2C, TTL, and SPI are well supported for most CPUs and have good software library support, your primary concern should be on the performance of your sensor.

The decision to use I2C, TTL, or SPI is a secondary consideration for satellite work because you will be customizing your satellite to your sensor in any case. A bad sensor cannot be fixed in software, whereas a very good sensor that uses a standard protocol (regardless of which protocol that is) will deliver good data.

11/Instrument Modes

For both reliability and flexibility, you will want to program your CPU to allow for different instrument modes. We've discussed the need to set lower and upper thresholds on your detector in order to accomodate the lowest signal you expect, and the highest signal you expect.

Dynamic Range

We're assuming, typically, that you've already chosen your image size or, if it's a non-imaging detector, it's effectively a single-pixel detector. Now you have to decide on an expected range that you wish to measure—an upper limit beyond which you do not expect to measure anything, and a lower limit that sets the minimum value you wish to measure. You also choose a desired accuracy for your measurement, which will determine your data resolution. The combination of these limits and the accuracy of your measurement will determine the dynamic range and resolution of your instrument.

Your dynamic range is simply (upper_limit – lower_limit), and your data resolution is dynamic_range/number_of_resolution_bins. Going to our earlier thermometer example, a thermometer that measures from 40°C to 70°C using 4 bits (0–15) has a dynamic range of 30°C, and a data resolution of 2 degrees (30°C/15), meaning any value read out is only accurate within 2 degrees.

The same thermometer, if reset to cover 10°C to 100°C, would have a larger dynamic range of 90°C but a worse resolution of only 6 degrees (90°C/15). This means you can detect higher and lower temperatures, but your numbers are only accurate +/–6°C.

Defining Multiple Instrument Modes

It was a very poor camera that, back in the day, had only one size FOV and required an exact amount of light to function. Instead, modern cameras let you zoom or go wide, let you shoot in bright or low light, and let you capture fast action or linger on a slow shot. About the only thing they don't do is let you play with different colors.

To use a given detector for a variety of potential situations, then, it is good to be able to reprogram your sensor's dynamic range and, optionally, number of bits of resolution. This lets you set up different instrument modes for different situations.

You will want to define your most common modes of operation. Then, you can write code that lets you alter a sensor's internal settings to tweak it for different situations.

Example Modes For a Magnetic Sensor

As an example, an I-CubeX magnetic sensor (*http://bit.ly/ZA4jyz*) defaults to operating with a fixed 7 bits of data (*brightness*) resolution, a 100ms timing resolution, and a base magnetic field (*spectral*) ranging from 0 to 16Gauss. Since 10 bits is 0–127, that nominally means that a magnetic resolution of 16Gauss/127 equals 0.125Gauss. However, that's because it is designed for use on Earth, where the field rarely drops below 0.3Gauss.

In space, we may have 0.06Gauss fields to measure. Keeping to 7 bits, we want a lower threshold of 0.06 and an upper of, perhaps, 0.3Gauss. We can do this by changing the dynamic range. We want a range from 0.3 to 0.06, or approximately 0.3Gauss, versus the as-shipped range of approximately 16Gauss.

However, it allows for a *gain* magnifier ranging from 1/64 to 2. Gain means that the *signal from the sensor* equals the *signal detected * gain*. So if our upper limit is a signal detected of 16Gauss and we use a gain of 1/64, our new upper limit is 16/64, or 0.25Gauss. Using gain like this, we get our desired dynamic range. While we're still only using 7 resolution bits for our data, all those bits are now sampling between our limits of 0–0.25Gauss and we are able to detect the weaker 0.06 fields we wish.

The sensor has a linear response of 1.5 to 10.5Gauss. Because we'll be below that threshold, we'll need to ground test to find the sensor's responseive curve, or wait until it is in orbit and hope to compare reference values obtained by other missions.

Cadence

You also may want the ability to change your timing or sampling rate. A sensor observing the Sun, for example, may only take one data frame per hour, but will increase this to one frame per minute during a flare event. Altering cadence (time between frames) allows for better data collection.

You can also schedule data collection cadence based on other criteria. Data management is primarily a task we'll cover in book 4 (Communications), as it is highly dependent on how many downlink opportunities you get. A big concern is what to do when your onboard memory is full: do you overwrite older data and only keep new, or stop taking data until you get free space? Again, there is no fixed answer; you must assess based on your mission goals.

In general, being able to change modes and allow for *triggered* special events greatly increases the odds that your detector will capture something interesting.

Triggering

One common sensor technique involves running the sensor at a low resolution with a wide dynamic range to constantly capture data. When something out of the ordinary happens—typically a very high data value—the CPU is set to automatically trigger a better data mode.

For example, this new mode may increase the resolution (and therefore the amount of bandwidth needed), so the event gets sampled more thoroughly than the quiescent (quieter) data. Alternatively, the new mode may take more data frames, to capture the timing and rise/fall of the event.

The new mode may simply adjust the dynamic range so that the detector dynamically adjusts to cover the appropriate range, with or without changing the actual resolution. It's also possible that the new mode could make all three changes.

Triggering and setting dynamic ranges is an excellent way to optimize the use of your sensor while minimizing bandwidth, but it does require clever programming. In contrast, a picosatellite with frequent ground contacts could rely on the ground operators to periodically command it to shift to a new instrument mode.

Instrument modes are usually preloaded as macros, rather than being defined on the fly. Telling the system to *load mode 4* is easier than addressing each individual register on a sensor. Therefore, when designing and programming your sensor, give strong consideration to having multiple modes with which it can collect data.

Automatic (unattended) triggering can be modeled in pseudocode as:

```
// reads out household temperature, also adapts to reading a tea kettle
// let our thermometer measure 40-60 C degrees unless things boil
lower_threshold_base = 20 // standard lower limit
lower_threshold_hot  = 99 // just below boiling

upper_threshold_base = 60 // standard upper limit
upper_threshold_hot = 200 // in case the water boils over

mode = "base"

while (signal = get_signal_from_detector()) do

  // change instrument mode
  if (mode == "base" and signal > upper_threshold_base) then
    // switch into 'hot mode'
     set_upper_limit(upper_threshold_hot)
```

```
    set_lower_limit(lower_threshold_hot)
  elseif (mode == "hot" and signal < lower_threshold_hot) then
    // switch back to 'room temperature mode'
    set_upper_limit(upper_threshold_base)
    set_lower_limit(lower_threshold_base)
  end if

  // and of course all your usual data collection goes here

end while
```

Non-Photon Detectors (And in situ)

It may seem we've been focusing on photon-based detectors, covering everything from radio to optical to high-energy photons. Fortunately, electric field or magnetic field detectors use the same math and same calculations. Thermometers, chemical detectors, impact detectors, Geiger counters, and any other detectors also use the same math.

These detectors tend to be called *in situ* (Latin for *in place*) detectors. This is because they do not capture a distant signal through a particular field of view, but instead capture the signal as it passes through the actual satellite/sensor.

A common picosatellite standard is a Hall effect sensor, which measures the strength of the magnetic field in a given direction. Clearly our same principals apply—it will have a given *brightness* range, which represents the strength of the magnetic field.

Most *in situ* detectors can be considered one-dimensional single-bit, or one-pixel in terms of *imaging*, because they return just one value at a time.

Similarly, they can be considered one-channel detectors in terms of spectral resolution, because they are measuring a single property.

Ergo, the bandwidth for an *in situ* detector is just 1 * 1 * (brightness resolution in bits) * (readout timing rate). This is the exact same calculation as before.

Some *in situ* detectors can be considered multi-channel because they bundle several similar detectors together. The I-CubeX "Orient3D," for example, determines your position in three dimensions using three separate Hall effect magnetic sensors: one aligned along the x-axis, one along the y-axis, and one along the z-axis. Therefore, it's just a three-channel, one-bit detector with given brightness resolution and timing resolution. Same logic, it's just a sensor!

CPUs

A satellite needs a brain. But which brain? IOS kits include the BasicX processor; I also tried the Arduino kit so beloved by DIY folks. Both are potentially flyable. Let's compare.

BasicX-24 (http://www.basicx.com/)
: 32K memory, requires 20mA plus up to 40mA I/O loads, operates at –40°C to +85°C. Programmed in BASIC via serial cable. Typically runs at 7.37MHz.

Arduino (http://arduino.cc)
: 16K memory, requires 50mA plus 40mA per I/O load, operates at –40°C to +125°C (estimate based on range at which the optional temperature sensor functions). Programmed in a C subset via USB port. Typically runs at 16MHz.

Raspberry Pi

The Raspberry Pi is an ultra-small PC that is often used as an embedded system or controller. It has USB, video, and keyboard connections, and runs Linux. Libraries are available to handle sensor protocols such as I2C, which you can then program in Python. It draws more power than an Arduino, BasicX-24, or PIC—about 3.5 Watts for the Pi, versus a half watt for the typical embedded controller. Due to its higher power consumption, we don't yet foresee it becoming the CPU for the power-starved picosatellites. But by all means, feel free to use it for early rapid sensor prototyping and development!

They Fight!

So we end up with a classic rocket science tradeoff. There's an easier to use, robust kit—the Arduino. And a more bare bones rig—the BasicX—that has better power usage but is a bit harder to hook up. Both are flyable. So do you go for ease of use, or performance?

If you want to go for performance, the lower power requirements of the BasicX are an overwhelming plus. As a computer scientist, I'm not worried about the ground issues—hooking it up, programming it, testing. This is a first launch, and I need to ensure everything is as *tight* as can be.

However, if my sensors were complex or I needed stronger software support, I'd be tempted to try the Arduino, based on the strong loyalty from the DIY community.

Both have a wide voltage rate (5–12V). The Arduino is 16MHz, and the BasicX-24 is nominally 7.37MHz but optimized to run its built-in BASIC faster.

The reason I limited it to those two choices is that I'm not up to creating a custom rig, so I went with one of the two most common kits. The BasicX-24 is a kit that comes with the IOS TubeSat, so it'll work with the PCB layouts they provide and is generally known to be *flight ready*. The Arduino is a known easy-to-use kit solution.

In order for me to switch from "IOS-approved BasicX-24" to "DIY-friendly Arduino," either there would have to be a problem with the BasicX-24, or the

Arduino would have to offer something much better than the BasicX-24. Since they are roughly comparable, and the BasicX-24 has a better power profile, while the Arduino has stronger programming support, the BasicX-24 wins this time.

However, this is one personal case. An axiom I teach is that you choose a given programming language for a task for any of three reasons:

1. It's the best for the task
2. You are familiar with it
3. It's what is already installed

You have to weigh each factor personally. Whether or not you know the language is an important factor, but it is also crucial to know whether or not it can support the sensors you are flying. There is no one correct choice. As more people fly Arduino-based CubeSats, that will clearly tilt the choice towards Arduino. But there are also people looking at using the Raspberry Pi, using ARM processors, and so on.

Your choice of CPU and development environment has to include support for your sensors, but also has to factor into your overall systems engineering plan.

Communications Limits

For my Project Calliope, my CPU has two tasks to perform.

1. Formatting the data (in my case, MIDI-format) into digital 128-byte AX.25 packets, and either sending them to the transmitter chip, or using an intermediate PIC to send them to the transmitter chip (basically, the role of a TNC)
2. Turning the transmitter on and off based on ground commands as per FCC regulations

While Book 4 will get into the hard details about communications, you need to know a bit about how you'll be getting your data down to Earth in order to effectively design your sensors. Most picosatellites will not have a data downlink to the ground that is available 24 hours a day, 7 days a week (24/7). Instead, you will need to select which data to keep, store it onboard, then downlink it when downlinks are available. Alternatively, you can simply make your satellite broadcast live data as it is taken during the communications downlinks. Both require you to assess how often you will have downlinks and how much bandwidth you will have, in order to determine the total size of the data you will receive.

For example, our low resolution CubeSat case in "Sampling and Bandwidth Calculations" on page 33 noted that packet radio allows up to 9600 bps, which translates to 1.2 KB/sec. If we assume you get one 10-minute contact pass per day (out of the five that pass over your radio command post), then you have 600*1.2KB, which equals a downlink budget of approximately 1MB of data per day. Therefore, whatever detectors we have, they cannot exceed that amount of useful data.

Amateur (HAM) Radio

Amateur satellites use shared spectrum with low-power transmitters. The satellites are moving quickly and most antennas need a directional fix on them to get the signal. The IARU tries to guarantee that everyone gets a chance to be heard, without one mission dominating the airwaves over the rest. That said, a satellite guy named Wes working on TubeSats just pointed us to an interesting coordination for packet data (*http://bit.ly/Yo6XtC*) from satellites, at, that I need to investigate further.

The FCC rules the spectrum, but the International Amateur Radio Union (IARU) is the entity that actually coordinates satellites. I need to file/coordinate with the IARU to use the amateur-band with my personal HAM callsign as the satellite's callsign. Like any regulation, there are many details.

The main IARU requirement is "Play nice, and be able to turn off your transmitter at a moment's notice." My approach will be to have the required *stop transmit* command, of course. To make this really work, however, I will also have all transmissions time out and shut down automatically after an appropriate interval (such as 10 minutes). Transmission only starts when a *start transmitting* command is uplinked, with possibly a few orbits that are prearranged to make the satellite automatically turn on based on a clock (if IARU allocates the time).

Since most HAMs use transceivers, which can receive and transmit, it is more fair to only broadcast when someone is willing and able to receive but it shouldn't reduce the ability to get Calliope data down. I need to do some minor orbit calculations to figure out the optimum window (est: 10 minutes), then apply to IARU to ask to share the frequency.

From a ground station listener point of view, when you wish to listen, you transmit the *start transmitting* command to Calliope, then it broadcasts for the set time (while you're in range) then shuts down. By definition, you can also shut it down earlier (since you're the one who sent the command and you are still in range), so it provides assurance that Calliope won't hog the spectrum.

This complies with the IARU requirements without needing me to guarantee 24/7 uplink to the satellite. Note that I have not yet contacted IARU, however, so this remains theoretical. I may be missing an FCC requirement somewhere. There could be roadblocks. IARU may say "No." Communication, both political and negotiated, is a component that is hard to parameterize early in a satellite build, but it becomes easier once your satellite is actually scheduled on a rocket launch manifest.

12/Off-the-Shelf Sensor Hardware

I can suggest three tactics for choosing your actual sensor hardware. The first is to conduct Web searches on sensors and pick a vendor. It takes a fair amount of research, but will yield a wider variety of sources.

The second is to explore what people are using for robotics and DIY works, and concentrate on understanding the sensors offered by those community-aimed vendors. For that sort of sensor work, you will often find good programming and installation support, since they are used to providing to hobbyists. Further, robotics work and satellite work are very similar.

The third is to really understand electronics, then buy directly from Digikey, Farnell, or other electronics shops. This requires knowledge, and is the most difficult path, yet it is also the one most likely to lead to the best performance.

Shopping

Often, companies that sell detectors for robots, drones, and satellites will indicate either the type of measurement (imaging, acceleration) or the wave length/data regime (light, IR, magnetic).

A lot of available sensors are not suitable for orbital use. Moisture sensors, sound sensors, and smoke sensors simply will not find anything measurable in orbit, unless they are measuring an item inside the picosatellite itself.

The main purpose of showing these lists is twofold. First, it's useful for giving you ideas for possible items to fly and missions to run.

Most importantly, it reflects the fact that there is no one schema for defining or selling sensors. One company's "magnetic field sensor" is another manufacturer's "Hall effect sensor" and a third's "biofield detector."

Here are several sample vendors' lists:

CuteDigi (*http://www.cutedigi.com/sensor*):

- Accelerometers
- Biometrics
- Current
- Flex/Force
- Gas Sensor
- ID
- Infrared
- Light/imaging
- Location/direction
- Magneto
- Proximity distance sensor
- Temperature

Trossen Robotics (*http://bit.ly/WnDZdD*):

- IR sensors
- Multi-axis accelerometers
- Multi-axis gyroscopes
- Compass sensors
- PIR motion sensors
- Object tracking & recognition
- Temperature and humidity measurement
- Ph, light, magnetic, and PSI sensors

Infusion Systems (*http://bit.ly/YbaqcK*):

- Air
- Bang
- BendMicro/BendMini/BendShort (flex of a material)
- Flash (visual light)
- GForce3D
- Hot (temperature)
- Light (visual light)
- Magnetic

- MoveAlong/MoveAround/MoveOn
- Orient/Orient3D
- Spin2D
- Stretch
- Turn
- Vibe

RobotShop (*http://www.robotshop.com/sensors.html*):

- High-end scanning lasers and obstacle detectors
- Infrared and light sensors
- Stretch and bend sensors
- Force sensors
- Accelerometers
- Gyroscopes
- Inertia measurement units
- Inclination & tilt sensors
- RFID
- Cameras and vision sensors
- Contact and proximity sensors
- Magnetic sensors/compass
- Current and voltage sensors
- Temperature and humidity sensors
- Thermal array sensors

Particle Damage

One concern is that orbit has a high flux of charged particles from the Sun, which can damage electronics. Radiation-hard parts are available, but cost more than off-the-shelf. Typical picosatellite mission lifetimes are short (less than 3 months), so I'm not worried about cumulative damage. I used to do radiation damage models back in school and it turns out that modern electronics are surprisingly robust on short time scales. We will primarily have single event upsets (SEPs) that scramble a sensor or computer, but since we don't need 100% uptime this shouldn't be a problem.

Be aware that many off-the-shelf parts may not exactly perform as their spec sheets suggest. One CubeSat team was using an imaging sensor that simply refused to perform as specified. It wasn't the sensor's fault so much as the vendor; their documentation did not match the actual sensor's interface and the interface code as provided was not robust. By that logic, you would be best served by choosing more than one sensor, then seeing which one performs best *for you* in *your lab* based on the skill sets of *your team*.

"Project Calliope" Sample Sensor Loadout

Given the fact that I suggest you try many sensors before settling on your final loadout, it's only fair I document my own efforts. I enjoyed testing the first batch of sensors I bought from ICube-X (two different Magnetics and a Hot). My final flight rig includes these sensors:

- (1)*Mag v2* for magnetic field strength
- (1)*Hot* temperature sensor
- (3)*Flash* and
- (1)*Light* (since the satellite will have an unpredictable spin)
- (1)*Orient3D* for magnetic field orientation

The function of the light sensors will be largely binary, to catch the bright/dark cycle as the satellite itself spins to face toward or away from the Sun, as well as catching the overall day/night cycle of each orbit. If there is a slight tumble to the satellite, all the better. These light sensors will provide a basic rhythm track.

Part of the reason I settled on this one vendor is that, by the time I'd spent hours figuring out their spec sheets and parameters, I simply did not want to take the time to learn yet another company's way of defining devices. Given that there is no one standard, just keeping up to date on the material can be difficult, unless you're an electrical engineer. So I chose this product because it's a good tool for the job, and I understand it, even though it may not technically be the absolute best tool available. I won't know what the best is until I encounter it. I chose ICube-X because, once I found it and bought my first set, I didn't need to search for something better/faster/cheaper.

Another reason for my choice was that I was able to get technical clarification on key parameters. For example, their techs suggested that their *Reach* detector, a capacitive charge detector, could potentially be used to detect high-energy particle events. In the end, due to time, I decided not to add it as a detector, but if my launch date gets pushed back further, I may revisit that decision and do a refit.

Sample Tech Advice

Axel Mulder of ICube-X wrote this bit of technical advice. I quote it in its entirely because this is exactly the sort of information you will need to understand when choosing your sensors. Remember, there is no one standard on sensors or capabilities, so you must really understand the extremes of a sensor's behavior before you can fly it with any reliability.

> I looked into the range and other specs of the Magnetic v2 sensor and updated the product's web page. The table on the web page can't show data below the listed 0.63 Gauss because there's too much background noise from the Earth's magnetic field. But I think you should be able to detect variations as low as 0.06Gauss if there's no other magnetic source around like the Earth's magnetic field. In fact, according to what I was able to calculate based on the few measurements we did some time ago with the magnet that we include, the lowest detectable value will be much lower than the 0.06Gauss you need, but the calibration curve will be non-linear, and you would need to create a test setup that eliminates the Earth's magnetic field. The noise such as from electrical noise in the circuit will be the limiting factor as far as I can see, and this should be quite a bit lower than your required 0.06Gauss.
>
> Axel Mulder

They also pointed out their sensors are only guaranteed to −40°C. Most electronics have trouble below −40°C, so just to support the computer and transmitter, I'm assuming my satellite will maintain a temperature above that.

That said, a metal plate in low Earth orbit will cycle from −170°C to 123°C, depending on time in sunlight. Since it's spinning, this range is fortunately smaller (as heat has time to distribute and dissipate), and with a 90-minute orbit, we should cycle through three ranges: too cold to register, transition regions where the sensor returns valid, slowly-changing data, and possible oversaturation at the high end. But I don't think we'll go much above 100°C, and I believe we'll usually be above −40°C. Obviously, thermal profiling is key.

ArduSat Sample Sensor Loadout

Early this March, ArduSat ("Your Arduino Experiment in Space") project announced their sensor loadout (*http://kck.st/14i3DVT*). Similar to Calliope, they are flying a magnetic field sensor, a brightness sensor, and a space-facing temperature sensor.

However, ArduSat adds greatly to the mix by tossing in a position gyro, an accelerometer for collecting attitude information, and a way to measure the satellite's internal temperature. They also add both an imaging camera and a spectrometer!

Specifically, ArduSat is using a Freescale MAG3110 three-axis magnetometer, an InvenSense ITG-3200 three-axis digital gyro, an Analog Devices ADXL345 three-axis accelerometer, a Melexis MLX90614 IR temperature sensor, a Texas Instruments TMO102 digital temperature sensor, a LND, Inc. Geiger counter, an Adafruit TSL2561 luminosity sensor for IR & visible light, a MySpectral Spectruino optical spectrometer, and a 1.3 megapixel optical camera. Except for the camera, all are low-bandwidth detectors.

Sensor Integration

True, I could just wire up a $10 Hall Effect sensor with an op amp, toss in some optical sensors, and call it a day. But where's the fun in that? The intent of Project Calliope was to show that someone without extraordinary skill in the art can build an effective musical satellite.

Using ICube-X sensors to create MIDI messages that feed into a BasicX board for transmission to Earth may not be the most technologically efficient or elegant approach, but it is the easiest. The problem becomes purely one of integrating systems, not of developing or heavy coding. To me, that shows that DIY picosatellites are feasible.

Art in Space

My own mission, Project Calliope, is to convert the ionosphere to music. Being an *art in space* project, it's perhaps more tolerant of issues like calibration errors and glitches. Primarily, I will be measuring deltas—relative changes in the ambient ionospheric magnetic field—rather than needing a precise field strength measurement.

In fact, glitches will add character to our derived music. Should we get, say, a solar storm, it'll be interesting to see how the sensors deal with it, either with saturation or with spurious signals.

Using delta measurements (how much a quantity changes, rather than what its absolute value is) is a prime example of a good use for picosatellites. The operational and programming needs are slightly simpler: you don't have to worry about response curves as much as you do in a mission requiring absolute calibration, and you can still return interesting data.

13/Committing, Freezing, Moving Forward

Every sensor is unique. There is no generic imaging sensor, universal CCD, standard temperature probe, or universal particle counter. You can no more swap out one brand of light sensor for a different make and model, than you could swap out a 100 Ohm 5% tolerant resistor for a 120 Ohm 10% tolerant resistor and expect your circuit to perform exactly as it did before without modification.

In practical terms, this means you should do a lot of initial shopping and testing. Then, once you've prototyped your sensor design and made your decision on what to fly, you lock yourself in to that particular manufacturer, make, and model.

The main reasons that you cannot just swap around sensors include differences in resolution, accuracy, and data range. In addition, a new sensor is likely to require a different wiring scheme. Finally, a new sensor is likely to require you to alter the computer programming needed to set up and read that particular sensor.

If a more advanced piece of sensor hardware becomes available later in your build cycle, you will have to evaluate the potentially significant extra time it will take to interface and parameterize the new sensor. Sometimes it may be worth improving your hardware; in other cases that extra time might be better spent making your existing hardware more robust and space-worthy.

Part of this arises from the precision that you require. A component that is both generic and has a wide tolerance—a standard resistor or a piece of wire—can be easily improved if desired. A more fundamental component like your choice of underlying CPU, your battery choice, or your sensor design, should not be changed lightly since change adds risk. Therefore, this risk must be balanced against the time cost and the potential benefit.

Buy Many

Most sensors do not perform exactly to their specifications.

This is not necessarily a bad thing. For a given set of presumed identical sensors, one may perform better than the others, while another might be flaky. Therefore, you should always buy three or more of a key component and test it to be sure you are using the one that—through the variance of manufacturing tolerances and the whims of Murphy's Law—empirically performs best on your lab bench.

Lego-Style Versus Custom Shop

Should you go with pre-made sensors, with smart sensors, or with custom circuits you designed? There are several schools of thought on building a CubeSat or other picosatellite. Let's contrast what we call "Lego style" with what we'll dub the "Custom Shop" approach.

Lego style involves using the easiest, rather than the most efficient, parts and tools to create your satellite. This is the kit-bashing or Lego bricks approach. You have several generic pieces, and you put them together to make what you want. The final end product may be a bit square and clunky, but the advantage is that you were able to quickly build and test.

The justification is time-based and presumes that, although small, the size, weight, and power margins of a CubeSat (relative to the desired payload) are sufficient enough that more generalized, but less efficient and less optimized, parts will suffice. The time saved in using openly available general solutions such as Arduino or BasicX-24 computers on a chip, or existing I2C sensors, outweighs the fact that they are not optimized for the desired space mission. The general kit solutions come with multiple hardware input/output channels and easy programming interfaces, so a team can quickly build and test its payload with minimal electrical engineering required.

Under the Lego style, almost any "brick" is good enough to begin prototype builds. If a given design ends up not working—if an Arduino, or an off-the-shelf imaging diode, or a standard batch of solar cells underperforms—you can simply swap in a different part and try again.

Lego style design work is at a higher level of abstraction because each component (power, transmitter, CPU, instrumentation) is isolated and using a well-defined protocol (USB, I2C, and so on). The components are loosely coupled and swappable.

The second approach, Custom Shop, argues that part selection and choice of components should be an intrinsic part of the design, not only from a performance and efficiency point of view, but also from a reliability perspective. Under this argument, using a specialized component such as a PIC (Peripheral Interface Controller), coupled with the exact number of ADCs (Analog-Digital Converters) needed, will result in a specialized hardware design that is electronically more simple. We call this the Custom Shop approach because you are specifically arguing that off-the-rack won't fit.

Justification for Custom Shop is that time spent early in parts selection and design, and the additional time cost for integration needs, is worth it when you get to testing and deploying your spacecraft. The higher reliability of the well-thought-out solution will result in fewer failures down the line, in part because your components are more tightly integrated.

Custom Shop also requires more expertise in electrical engineering, since the team is not just using defined components but is required to choose and design the system at a lower level of abstraction, both choosing specific circuit parts and dealing with the hardware directly rather than through existing modules or libraries. To some degree, your ability to delve deeply into design is always going to be a function of your talent pool size. If you have a large enough team with the necessary skills, certainly a higher level of reliability by using specialized solutions is worth considering.

Both approaches are valid, with strengths in different contexts. The primary justification for Lego style for CubeSats is that most CubeSat missions are lower-cost and higher-risk than a typical satellite. Further, most CubeSat teams work under a shorter development schedule than an industrial project team. By definition, if you're an undergraduate who wants to see your satellite fly, you're restricted to a two- to three-year development cycle.

However, lower cost is relative; a CubeSat budget of $100K is not unusual. Further, you only get very few chances for your CubeSat to work; a lab with consecutive failures is not going to be high on the queue for funding or launch opportunities. Finally, and most importantly, space is unforgiving, and if your satellite does fail, you can't go up and fix it.

To recap, Lego style allows for rapid design and development at the cost of using potentially less reliable parts and more of your system resources under a design that is not optimized. Custom Shop creates a project with higher reliability, but requires more expertise and more time to develop. Choosing which is right for your picosatellite team is the first challenge you will face, but remember...the clock is ticking.

A/Exercises

To provide some armchair satellite-building practice, here are two design exercises that ask you to invent a satellite. These can be done for fun, used at a Maker Faire (*http://makerfaire.com*) or an Unconference, attempted in a classroom setting, or brought out to impress non-satellite-builders.

The point is not whether you can build, but whether you can conceive and outline an idea that is worth building in the first place.

JWST Build-a-Satellite

Within 20 minutes, can you use a web tool to build a viable satellite?

1. Go to the JWST Build a Satellite page (*http://www.jwst.nasa.gov/build.html*) and launch the game.
2. Choose "Level 3."
3. Create a mission to explore Galaxies that (at least) uses a high energy band.
4. Try to successfully make a satellite. Write down your intermediate decisions (what wavelengths, what instruments, what optics). Be meaningful in your choices. Could you explain *how* and *why* you made your choices to a layman (or a spouse, or a roommate, or a barista)?
5. If it takes more than three tries and you still haven't made a successful satellite, feel free to drop to Level 2. Work up to Level 3.

Solve a Decadal Problem for All of Humanity

Choose one of the decadal goals for Earth observing, heliophysics, astronomy, or planetary science, and design a mission concept to fulfill that task using a small satellite platform—NASA SMEX or smaller.

Invent your satellite and make a five-minute pitch that you would present to NASA to ask for funding. Limit yourself to a satellite with one or two (at most) instruments. Here are some decadal reference links:

- *http://www.spacepolicyonline.com/national-research-council#decadal*
- *http://decadal.gsfc.nasa.gov/about.html*
- *http://science.nasa.gov/about-us/science-strategy/decadal-surveys/*
- *http://solarsystem.nasa.gov/2013decadal/*
- *http://science.nasa.gov/heliophysics/decadal-surveys/*
- *http://science.nasa.gov/earth-science/decadal-surveys/*

One example of a decadal goal, from Earth observing, might be:

> Changing ice sheets and sea level. Will there be catastrophic collapse of the major ice sheets, including those of Greenland and the West Antarctic and, if so, how rapidly will this occur? What will be the time patterns of sea-level rise as a result?

A good pitch might include:

1. The mission summary chart we presented at the start of this book (type/wavelength/goal/who/orbit)
2. History of any past missions that tackled this
3. List of desired instrument loadout: what instrument types and what they each measure plus whether or not it needs focusing optics
4. Resolution range per detector (spatial, spectral, timing, brightness)
5. Cost estimate, based on comparison/analogy to similar missions

To evaluate a good pitch, consider whether:

1. Your goal and satellite are plausible.
2. Your approach clearly seems to be the right approach for the task.

This is the skill of both business and academic proposals, where you must not only convince the audience that you are the right person for the task, but also that the task itself is worth doing!

About the Author

"Sandy" Antunes is an astrophysicist who turned to science writing upon realizing that having a desire to understand the universe doesn't mean you have to be the one to discover everything personally. There's a lot of excellent science out there, and Sandy enjoys bringing it to the world's attention. Sandy recently achieved a professorship at Capitol College's Astronautical Engineering department, which he credits to his NASA work, his solo build of the Project Calliope picosatellite, and his writing for Science 2.0 and O'Reilly Media.

DIY Instruments for Amateur Space is the third book in the four-book DIY Satellite series. The first, *DIY Satellite Platforms*, covers the basics in designing and building your own satellite; the second, *Surviving Orbit the DIY Way*, is about the environment of space and how to create a survivable design. Book 4 will provide information on how to communicate with and operate your picosatellite once it is safely up in space.

The cover and body font is BentonSans, the heading font is Serifa, and the code font is Bitstreams Vera Sans Mono.

CPSIA information can be obtained at www.ICGtesting.com
Printed in the USA
LVOW04s2243120415

434315LV00012B/154/P